Biomass: Feedstocks and Processing Techniques

Biomass: Feedstocks and Processing Techniques

Josh Olson

Larsen & Keller
www.larsen-keller.com

Biomass: Feedstocks and Processing Techniques
Josh Olson
ISBN: 978-1-64172-414-2 (Hardback)

Larsen & Keller
Published by Larsen and Keller Education,
5 Penn Plaza,
19th Floor,
New York, NY 10001, USA

Cataloging-in-Publication Data

Biomass : feedstocks and processing techniques / Josh Olson.
p. cm.
Includes bibliographical references and index.
ISBN 978-1-64172-414-2
1. Biomass energy. 2. Biomass. 3. Feedstock. 4. Biotechnology. I. Olson, Josh.
TP339 .B563 2020
662.88--dc23

Printed and bound in China.

For more information regarding Larsen and Keller Education and its products, please visit the publisher's website www.larsen-keller.com

TABLE OF CONTENTS

PREFACE

This book aims to help a broader range of students by exploring a wide variety of significant topics related to this discipline. It will help students in achieving a higher level of understanding of the subject and excel in their respective fields. This book would not have been possible without the unwavered support of my senior professors who took out the time to provide me feedback and help me with the process. I would also like to thank my family for their patience and support.

The plant or animal material that is used as raw material for the production of energy and heat as well as for the manufacturing of various industrial products is known as biomass. It includes energy crops such as miscanthus and switchgrass, forest residues, waste from crops, animal waste, and human waste from sewage plants. It is classified as renewable or green energy because it recycles the carbon dioxide back into new crops through photosynthesis. There are numerous methods of converting biomass into fuel. These methods are broadly classified into thermal conversion, chemical conversion, biochemical conversion and electrochemical conversion. This book unfolds the innovative aspects of biomass. The ever growing need of advanced technology is the reason that has fueled the research in this field in recent times. The coherent flow of topics, student-friendly language and extensive use of examples make this book an invaluable source of knowledge.

A brief overview of the book contents is provided below:

Chapter - Introduction

The animal or plant material used for energy production as well as in various industrial processes is known as biomass. This is an introductory chapter which will introduce briefly all the significant aspects of biomass as well as its resources.

Chapter - Significant Aspects of Biomass

Some of the major concepts related to biomass include bioconversion, energy forestry, bio-energy with energy capture and storage, ethanol fuel energy balance, and biomass allocation. The chapter closely examines these key concepts to provide an extensive understanding of the subject.

Chapter - Biomass Feedstocks

Any renewable and biological material that can be used directly as a fuel or converted to another form of fuel is known as a feedstock. Biomass feedstocks are the plants and materials used to derive fuels like ethanol, butanol, biodiesel, etc. The topics elaborated in this chapter will help in gaining a better perspective about different types of biomass feedstocks such as lignocellulosic biomass, miscanthus giganteus, landfill gas and jatropha curcas.

Chapter – Biomass Conversion Techniques

Some of the techniques used for the conversion of biomass are thermal conversion, bioconversion, chemical conversion, biochemical conversion, thermal depolymerization and electrochemical conversion. The diverse applications of these biomass conversion techniques have been thoroughly discussed in this chapter.

Chapter – Biofuel

Biofuel is a fuel which is produced from biomass through modern processes, instead of geologic processes. The common types of biofuels include bioethanol, biohydrogen, biogas, biobutanol and biodiesel. This chapter has been carefully written to provide an easy understanding of these types of biofuels.

Josh Olson

Introduction

CHAPTER 1

The animal or plant material used for energy production as well as in various industrial processes is known as biomass. This is an introductory chapter which will introduce briefly all the significant aspects of biomass as well as its resources.

BIOMASS

Biomass is organic material that comes from plants and animals, and it is a renewable source of energy. It contains stored energy from the sun. Plants absorb the sun's energy in a process called photosynthesis. When biomass is burned, the chemical energy in biomass is released as heat. Biomass can be burned directly or converted to liquid biofuels or biogas that can be burned as fuels.

Examples of biomass and their uses for energy:

- Wood and wood processing wastes—burned to heat buildings, to produce process heat in industry, and to generate electricity.
- Agricultural crops and waste materials—burned as a fuel or converted to liquid biofuels.
- Food, yard, and wood waste in garbage—burned to generate electricity in power plants or converted to biogas in landfills.
- Animal manure and human sewage—converted to biogas, which can be burned as a fuel.

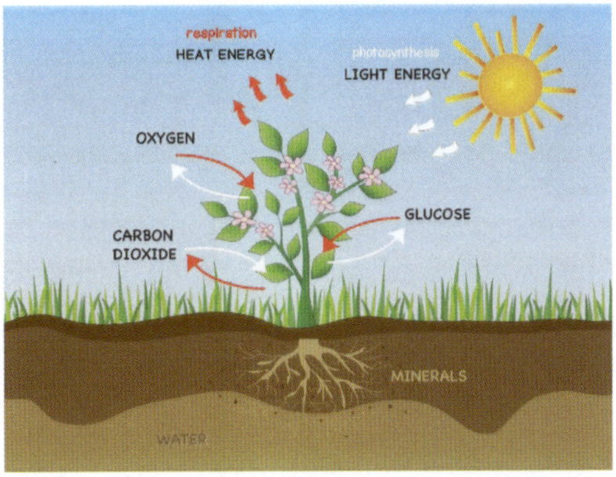

Converting Biomass to Energy

Solid biomass, such as wood and garbage, can be burned directly to produce heat. Biomass can also be converted into a gas called biogas or into liquid biofuels such as ethanol and biodiesel. These fuels can then be burned for energy.

Biogas forms when paper, food scraps, and yard waste decompose in landfills, and it can be produced by processing sewage and animal manure in special vessels called digesters.

Ethanol is made from crops such as corn and sugar cane that are fermented to produce fuel ethanol for use in vehicles. Biodiesel is produced from vegetable oils and animal fats and can be used in vehicles and as heating oil.

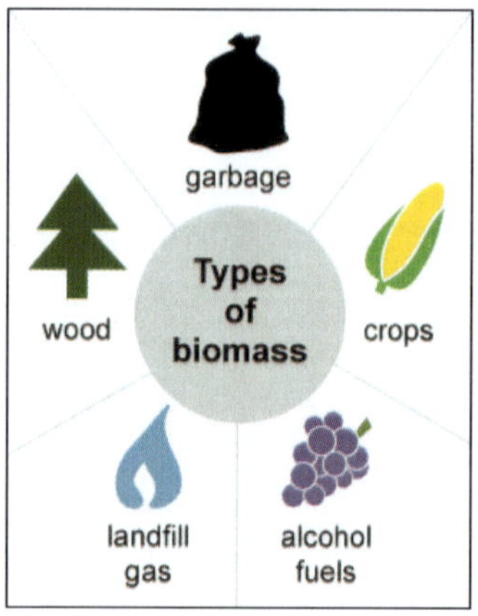

Biomass as Fuel

Biomass fuels provided about 5% of total primary energy use in the United States in 2018. Of that 5%, about 46% was from biofuels (mainly ethanol), 44% was from wood and wood-derived biomass, and 10% was from the biomass in municipal waste. Researchers are trying to develop ways to use more biomass for fuel.

Biomass for Electricity Generation

Biomass is used for facility heating, electric power generation, and combined heat and power. The term biomass encompasses a large variety of materials, including wood from various sources, agricultural residues, and animal and human waste.

Biomass can be converted into electric power through several methods. The most common is direct combustion of biomass material, such as agricultural waste or woody materials. Other options include gasification, pyrolysis, and anaerobic digestion. Gasification produces a synthesis gas with usable energy content by heating the biomass with less oxygen than needed for complete combustion. Pyrolysis yields bio-oil by rapidly heating the biomass in the absence of oxygen. Anaerobic digestion produces a renewable natural gas when organic matter is decomposed by bacteria in the absence of oxygen.

Different methods work bet with different types of biomass. Typically, woody biomass such as wood chips, pellets, and sawdust are combusted or gasified to generate electricity. Corn stover and wheat straw residues are baled for combustion or converted into a gas using an anaerobic digester. Very wet wastes, like animal and human wastes, are converted into a medium-energy content gas in an anaerobic digester. In addition, most other types of biomass can be converted into bio-oil through pyrolysis, which can then be used in boilers and furnaces.

Compared to many other renewable energy options, biomass has the advantage of dispatchability, meaning it is controllable and available when needed, similar to fossil fuel electric generation systems. The disadvantage of biomass for electricity generation, however, is that the fuel needs to be procured, delivered, stored, and paid for. Also, biomass combustion produces emissions, which must be carefully monitored and controlled to comply with regulations.

Most biopower plants use direct-fired combustion systems. They burn biomass directly to produce high-pressure steam that drives a turbine generator to make electricity. In some biomass industries, the extracted or spent steam from the power plant is also used for manufacturing processes or to heat buildings. These combined heat and power (CHP) systems greatly increase overall energy efficiency to approximately 80%, from the standard biomass electricity-only systems with efficiencies of approximately 20%. Seasonal heating requirements will impact the CHP system efficiency.

A simple biomass electric generation system is made up of several key components. For a steam cycle, this includes some combination of the following items:

- Fuel storage and handling equipment,
- Combustor/furnace,
- Boiler,
- Pumps,
- Fans,
- Steam turbine,
- Generator,
- Condenser,
- Cooling tower,
- Exhaust/emissions controls,
- System controls (automated).

Direct combustion systems feed a biomass feedstock into a combustor or furnace, where the biomass is burned with excess air to heat water in a boiler to create steam. Instead of direct combustion, some developing technologies gasify the biomass to produce a combustible gas, and others produce pyrolysis oils that can be used to replace liquid fuels. Boiler fuel can include wood chips, pellets, sawdust, or bio-oil. Steam from the boiler is then expanded through a steam turbine, which spins to run a generator and produce electricity.

In general, all biomass systems require fuel storage space and some type of fuel handling equipment and controls. A system using wood chips, sawdust, or pellets typically use a bunker or silo for short-term storage and an outside fuel yard for larger storage. An automated control system conveys the fuel from the outside storage area using some combination of cranes, stackers, reclaimers, front-end loaders, belts, augers, and pneumatic transport. Manual equipment, like front loaders, can be used to transfer biomass from the piles to the bunkers, but this method will incur significant cost in labor and equipment operations and maintenance (O&M). A less labor-intensive option is to use automated stackers to build the piles and reclaimers to move chips from the piles to the chip bunker or silo.

Wood chip-fired electric power systems typically use one dry ton per megawatt-hour of electricity production. This approximation is typical of wet wood systems and is useful for a first approximation of fuel use and storage requirements but the actual value will vary with system efficiency. For comparison, this is equivalent to 20% HHV efficiency with 17 MMBtu/ton wood. Most wood chips produced from green lumber will have a moisture content of 40% to 55%, wet basis, which means that a ton of green fuel will contain 800 to 1,100 pounds of water. This water will reduce the recoverable energy content of the material, and reduce the efficiency of the boiler, as the water must be evaporated in the first stages of combustion.

The biggest problems with biomass-fired plants are in handling and pre-processing the fuel. This is the case with both small grate-fired plants and large suspension-fired plants. Drying the biomass before combusting or gasifying it improves the overall process efficiency, but may not be economically viable in many cases.

Exhaust systems are used to vent combustion by-products to the environment. Emission controls might include a cyclone or multi-cyclone, a baghouse, or an electrostatic precipitator. The primary function of all of the equipment listed is particulate matter control, and is listed in order of increasing capital cost and effectiveness. Cyclones and multi-cyclones can be used as pre-collectors to remove larger particles upstream of a baghouse (fabric filter) or electrostatic precipitator.

In addition, emission controls for unburned hydrocarbons, oxides of nitrogen, and sulfur might be required, depending on fuel properties and local, state, and Federal regulations.

Working Principle

In a direct combustion system, biomass is burned in a combustor or furnace to generate hot gas, which is fed into a boiler to generate steam, which is expanded through a steam turbine or steam engine to produce mechanical or electrical energy.

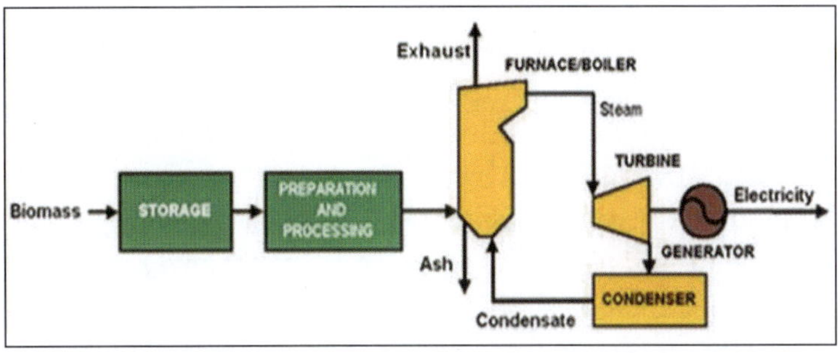

In a direct combustion system, processed biomass is the boiler fuel that produces steam to operate a steam turbine and generator to make electricity.

Types and Costs of Technology

There are numerous companies, primarily in Europe, that sell small-scale engines and combined heat and power systems that can run on biogas, natural gas, or propane. Some of these systems are available in the United States, with outputs from about 2 kilowatts (kW), and approximately 20,000 British thermal units (Btu) per hour of heat, to several megawatts (MW). In addition, small-scale (100 to 1,500 kW) steam engine/ gen-sets and steam turbines (100 to 5,000 kW) that are fueled by solid biomass are currently available in Europe.

In the United States, direct combustion is the most common method of producing heat from biomass. Small-scale biomass electric plants have installed costs of $3,000 to $4,000 per kW, and a levelized cost of energy of $0.8 to $0.15 per kilowatt hour (kWh).

The two principal types of chip-fired direct combustion systems are stationary- and traveling-grate combustors, otherwise known as fixed-bed stokers and atmospheric fluidized-bed combustors.

Fixed-bed Systems

There are various configurations of fixed-bed systems, but the common characteristic is that fuel is delivered in some manner onto a grate where it reacts with oxygen in the air. This is an exothermic reaction that produces very hot gases and generates steam in the heat exchanger section of the boiler.

Fluidized-bed Systems

In either a circulating fluidized-bed or bubbling fluidized-bed system, the biomass is burned in a hot bed of suspended, incombustible particles, such as sand. Compared to grate combustors, fluidized-bed systems generally produce more complete carbon conversion, resulting in reduced emissions and improved system efficiency. In addition, fluidized-bed boilers can use a wider range of feedstocks. Furthermore, fluidized-bed systems have a higher parasitic electric load than fixed-bed systems due to increased fan power requirements.

Biomass Gasification Systems

Although less common, biomass gasification systems are similar to combustion systems, except that the quantity of air is limited, and thus produce a clean fuel gas with a usable heating value in contrast to combustion, in which the off gas does not have a usable heating value. Clean fuel gas provides the ability to power many different kinds of gas-based prime movers, such as internal combustion engines, Stirling engines, thermo electric generators, solid oxide fuel cells, and micro-turbines.

Small, modular biopower system by Community Power Corporation.

The efficiency of a direct combustion or biomass gasification system is influenced by a number of factors, including biomass moisture content, combustion air distribution and amounts (excess air), operating temperature and pressure, and flue gas (exhaust) temperature.

Application

The type of system best suited to a particular application depends on many factors, including availability and cost of each type of biomass (e.g. chip, pellet, or logs), competing fuel cost (e.g. fuel oil and natural gas), peak and annual electrical loads and costs, building size and type, space availability, operation and maintenance staff availability, and local emissions regulations.

Projects that can make use of both electricity production and thermal energy from biomass energy systems are often the most cost effective. If a location has predictable access to year-round, affordable biomass resources, then some combination of biomass heat and electricity production may be a good option. Transportation of fuel accounts for a significant amount of its cost, so resources should ideally be available from local sources. In addition, a facility will typically need to store biomass feedstocks on-site, so site access and storage are factors to consider.

As with any on-site electricity technology, the electricity generating system will need to be interconnected to the utility grid. The rules for interconnection may be different if the system is a combined heat and power system instead of only for electricity production. The ability to take advantage of net metering may also be crucial to system economics.

Bioliquids

Bioliquids are liquid fuels made from biomass for energy purposes other than transport (i.e. heating and electricity). Bioliquids are usually made from virgin or used vegetable and seed oils, like palm or soya oil. These oils are burned in a power station to create heat, which can then be used to warm homes or boil water to make steam. This steam can then be used to drive a turbine to generate electricity.

Rudolf Diesel's first public exhibition of the internal combustion engine, that was to later bear his name, ran on peanut oil.

Bioliquid Production and Use

Bioliquids have been used for many years to provide heat for homes on a small scale but now big energy providers are looking at their use on a larger scale.

A controversial plant in Bristol (UK) was recently given the go ahead despite receiving several hundred complaints. The plant will be built and operated by W4B and provide enough power for 25,000 homes.

Advantages

Bioliquids have several key advantages over other sources of renewable energy:

- Bioliquids have a high energy density.
- The technology is well established, having been used for many years.
- Can be used on demand, reacting quickly to changes in demand for power.
- Can help reduce dependency on foreign oil.
- Reduces the green house gas emissions.

Disadvantages

Many of the same problems that affect biofuels also affect bioliquids and there are various social, economic, environmental and technical issues, which have been discussed in the popular media and scientific journals. These include: the effect of moderating oil prices, the "food vs fuel" debate, poverty reduction potential, carbon emissions levels, sustainable biofuel production, deforestation and soil erosion, loss of biodiversity, impact on water resources, as well as energy balance and efficiency.

Biomass to Liquid

Biomass to liquid (BtL or BMtL) is a multi-step process of producing synthetic hydrocarbon fuels made from biomass via a thermochemical route. Such a fuel has been called grassoline.

Main Processes

According to a study done by the U.S. Department of Agriculture and the Department of Energy, the United States can produce at least 1.3 billion tons of cellulosic biomass each year without decreasing the amount of biomass needed for our food, animal feed, or exports. These are following processes:

1. Fischer–Tropsch Process:

The Fischer–Tropsch process is used to produce synfuels from gasified biomass. Carbonaceous material is gasified and the gas is processed to make purified syngas (a mixture of carbon monoxide and hydrogen). The Fischer–Tropsch polymerizes syngas into diesel-range hydrocarbons. While biodiesel and bio-ethanol production so far only use parts of a plant, i.e. oil, sugar, starch or cellulose, BtL production can gasify and utilize the entire plant.

2. Flash Pyrolysis:

Flash pyrolysis—producing bio-oil (pyrolysis oil), char and gas at temperatures between 350–550 °C and residence times < 1 second (also called anhydrous pyrolysis).

3. Catalytic Fast Pyrolysis:

Catalytic fast pyrolysis is a fast process in which the cellulose is broken down to a liquid biofuel. In this approach the cellulose is heated to 500 degrees Celsius in less than one second in a chamber to break apart the oxygen molecules. The catalyst forms chemical reactions that remove oxygen bonds and form carbon rings. After the reaction takes place gasoline is formed along with water, carbon dioxide, and carbon monoxide.

4. AFEX Treatment:

The Ammonia Fiber Expansion (AFEX) pre-treat process, hot concentrated 15 M ammonia is used to break down sugar molecules, cellulose and hemicellulose significantly more efficiently than enzymes. After, the rapid pressure release cools and ends the treatment. The result is minor biomass degradation with high yields. The process was patented by Bruce Dale, Michigan State University professor. AFEX is generally done in one step, making it more efficient than other processes.

5. AFEX Process Conditions:

- Pressure: 20–30 atm,
- Temperature: 70–140 °C,
- Residence time: 5–10 minutes,
- Ammonia: biomass loading: 0.3 – 2.0 to 1 w/w,
- Water: biomass content: 0.2 – 2.5 to 1 w/w.

6. Catalytic Depolymerization:

Catalytic depolymerization is the use of heat and catalysts to separate usable diesel fuel from hydrocarbon wastes.

7. Regional Biomass Processing Center:

Regional Biomass Processing Center is a conceptual place where the AFEX treated biomass can go to biorefineries, farms and forests, and animal feeders. This will improve the value of cellulosic biomass for animals and biofuel production. This will reduce the density of the biomass for easier transport, simplify contract issues, and increase the land use for biofuels

The process uses the whole plant to improve the carbon dioxide balance and increase yield.

Potential Energy Grasses

Plant material is cheaper than oil on both energy and mass basis and certain plant material have potential to be energy grasses.

Switchgrass

Switchgrass is a bunch grass native to North America that grow naturally under warm weather with wide adaptation capability and easy germination, allowing the switchgrass to grow quicker; however, it has a low relative yield compared to other energy crops.

Sorghum

Sorghum are cultivated in warmer climates, mostly in the tropical regions. *Sorghum* has the potential to be an energy grass because it requires low water usage and can make a large yield. *Sorghum*, however, has an annual cultivation and is difficult to establish to an area and requires a lot of input from fertilizers and pesticides.

Miscanthus

Miscanthus are native to the tropical regions of Africa and Southern Asia. Miscanthus can grow up to 3.5 meters and has been trialed as a biofuel since the 1980s. The benefits of using miscanthus is it can live more than two years and requires low input eliminating the need for extra irrigation, fertilizer and pesticides. The problems with miscanthus arise from the time it takes to establish to an area.

Grassoline

Grassoline is a term coined in 1991 by Matthew Scoggins, a graduate student of Bruce Dale, to capture the idea of taking plant material and converting it into oil.

Cost of Change

The cost for petroleum to change to grassoline would depend on how fast the use of grassoline grows. Change will also be needed in automobiles to be compatible with grassoline. UC Berkeley's Somerville (professor of alternative energy) estimates that a large investment over $325 billion will be needed to build biofactories that can produce the 65 billion gallons of biofuel needed to meet 2030 national goals.

Cellulosic Sugars

Cellulosic sugars are derived from non-food biomass (e.g. wood, agricultural residues, municipal solid waste). The biomass is primarily composed of carbohydrate polymers cellulose, hemicellulose, and an aromatic polymer (lignin). The hemicellulose is a polymer of mainly five-carbon sugars $C_5H_{10}O_5$ (xylose). and the cellulose is a polymer of six-carbon sugar $C_6H_{12}O_6$ (glucose). Cellulose fibers are considered to be a plant's structural building blocks and are tightly bound to lignin, but the biomass can be deconstructed using Acid hydrolysis, enzymatic hydrolysis, organosolv dissolution, autohydrolysis or supercritical hydrolysis.

Biomass (cellulose, hemicellulose and lignocellulose) contain vast amounts of fermentable sugars. These sugars may be produced from a wide variety of feedstocks and can be converted into a multitude of biochemical, biofuel, and polymer products by either biological or chemical routes.

Industrial Use

In January 2012, BASF invested in Pennsylvania-based Renmatix to produce low-cost, large volume quantities of industrial sugar from lignocellulosic biomass (wood, cane bagasse or straw). Renmatix is currently the only commercial player utilizing supercritical hydrolysis as a route to cellulosic sugar production.

Renmatix is working with multiple partners on development of commercial scale facilities with the capability to produce more than 100,000 tons of cellulosic sugars annually. The company has a world-class technical center in Pennsylvania and production operations at the Integrated Plantrose Complex (IPC) in Kennesaw, Georgia and the Feedstock Processing Facility (FPF) in Rome, New York.

In June 2013, Renmatix also entered a joint development agreement (JDA) with UPM, a Finnish pulp, paper and timber manufacturer, to convert woody biomass into low-cost sugar intermediates for subsequent downstream processing into biochemicals.

In December 2013, Renmatix and Virent announced a strategic collaboration to convert affordable cellulosic sugars to renewable chemicals and bio-based packaging materials.

In March 2015, French Energy Group, Total S.A. entered a joint development agreement (JDA) with Renmatix to use the Plantrose technology to extract second-generation sugars from biomass and develop sustainable and profitable biomolecules for products of interest.

Table: Biotechnology penetration in the chemical industry.

Year	Value	Penetration
2000 (actual)	$67 billion	5.3%
2005 (actual)	$98 billion	6.7%
2010 (forecast)	$159 billion	9.6%
2025 (projection)	$1000 billion	33%

Table: World biobased market penetration 2010-2025.

Chemical Sector	2010	2015
Commodity Chemicals	1-2%	6-10%
Specialty Chemicals	20-25%	45-50%
Fine Chemicals	20-25%	45-50%
Polymers	5-10%	10-20%

In the first quarter of 2013, American Process Inc. announced the start-up of cellulosic sugar production using their patented AVAP® technology at their demonstration plant in Thomaston, GA. The AVAP process uses ethanol and sulfur dioxide (SO2) to fractionate biomass into its pure components: cellulose, hemicellulose sugars and lignin. In early 2013 GranBio, a Brazilian pioneer in biofuels and biochemicals, completed the acquisition of an equity investment in API. Since that time, API has fermented both five-carbon and six-carbon sugars into high value bio-chemicals and biofuels in partnership with fermentation companies throughout the world.

Applications

Cellulosic sugars are used as renewable resources for biochemical and biofuels industries and can be used to produce intermediates by fermentative processes. The availability of industrial sugars from renewable resources, in sufficient quantities and at a favorable cost enables the products to be cost-competitive to fossil fuel based products.

A 2012 study by Nexant estimates that in the future, it will be possible and potentially economically viable to produce any type of sugar-based chemical product from biomass due to developments in cellulosic processing.

Advantages of Biomass

Biomass has been in existence before people started talking about renewable energy sources and energy efficiency. This long term use would not be possible if biomass didn't have any benefit. As most governments are leading campaigns to find alternative sources of fuel to fossils, biomass energy continues to make headlines as a possible alternative. Here are the advantages that make biomass energy a perfect alternative to fossil fuels:

1. It's a renewable form of energy:

Biomass energy is considered a renewable form of energy because the organic materials used to produce it are never-ending. The organic materials including wood, crop waste, garbage, sewage sludge, and manure are continually produced by society. In a nutshell, regrowth of these organic materials supports the fact that biomass is renewable.

2. It's carbon neutral:

We all know that release of vast amounts of carbon contributes greatly to climate change. Biomass energy takes care of this since it is a natural part of the carbon cycle as opposed to fossil-based sources of fuel such as oil, natural gas, and coal. Researchers say that the only carbon emitted to the environment from biomass fuels is the amount that was absorbed by plants in the course of their life cycle. In the process of replenishing the used plant materials, the new ones that spring up absorb equal quantity of carbon, hence, developing neutrality that witnesses no new carbon generated. This aspect renders biomass uniquely clean.

3. Widely available:

Just like sun and wind energy, biomass energy sources are bountiful in supply. You can find them virtually in every nook and cranny of the world. The fact that it's bountiful in supply means that we may never encounter problems that we are presently experiencing with fossil-based sources of fuel. Nonetheless, it's vital that we maintain the abundance of this natural resource by being responsible in its use.

4. It's cheaper compared to fossil fuels:

Producing biomass energy does not involve heavy capital outlay. Fossil fuels production, on the other hand, involves high upfront capital costs such as drilling to reach oil wells, constructing gas pipelines and collection of biomass fuel. The low cost resulting from the production of biomass fuel is passed on to customers. This means that customer's energy bills will not depend on aspects such as availability and knee jack decisions of energy production and supply firms. Low biomass cost makes this form of energy attractive to manufacturers and producers since they are able to generate higher profits from extremely low output.

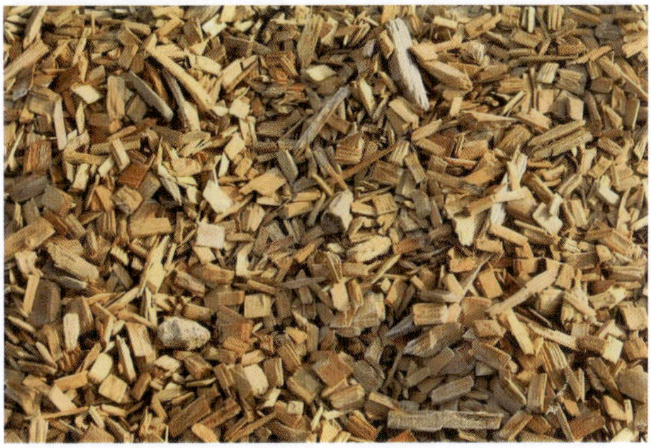

5. Minimizes overdependence on traditional electricity:

Virtually anyone can produce biomass energy because the raw materials are available everywhere. Traditional forms of electricity can sometimes be unreliable due to power outages. Biomass energy produced at home can be a good backup for traditional electricity.

6. Reduces amount of waste in landfills:

Most waste produced in homes is either plant matter or biodegradable. This kind of waste can be channeled to more profitable use. Biomass energy generation utilizes any waste that would have otherwise found way into landfills. This minimizes the impacts of waste in landfills to the environment. This impact may be compounded by contamination of local habitats and destruction of wildlife ecosystems. Minimized waste means reduction of land intended for landfills, hence, more space for human habitats.

7. Can be used to create different products:

Biomass energy is also versatile, as different forms of organic matter can be used to create different products. Ethanol and similar fuels can be made from corn and other crops. With so many living things on the planet, there is no limit to how many ways it can be found and used.

Disadvantages of Biomass

While the upsides to biomass energy are plenty, it's not exactly a perfect source of energy. Here are the downsides to biomass energy:

1. Not entirely clean:

Using animal and human waste to power engines may save on carbon dioxide emissions, but it increases methane gases, which are also harmful to the Earth's ozone layer. So really, we are no better off environmentally for using one or the other. And speaking of using waste products, there is the smell to consider. While it is not physically harmful, it is definitely unpleasant, and it can attract unwanted pests (rats, flies) and spread bacteria and infection.

2. Risk of deforestation:

Biomass energy sources are renewable, but they have to be utilized sustainably. Uncontrolled biomass production can result in deforestation. If deforestation is allowed to happen, scores of animal and bird species would be rendered homeless, not to mention the drought as a result. In fact, this is the main reason for slowing down the large scale use of biomass fuel. Governments feel replanting efforts may not match the rate of cutting down of trees.

3. Requires a great deal of water:

This is the most invisible disadvantage of biomass fuel. All plant matter need sufficient amount of water to get by, meaning water sources must be abundant. If enough water is not available, irrigation systems would have to be developed, which could prove to be costly. Irrigation may also limit the availability of water to humans and wildlife.

4. Inefficient as compared to fossil fuels:

Despite the fact that biomass energy is natural in many ways, it doesn't get close to fossil fuels in regards to efficiency. In fact, some renewable sources of energy like biofuels are fortified with fossil fuels to increase their efficiency.

BIOMASS RESOURCES

Biomass resources that are available on a renewable basis and are used either directly as a fuel or converted to another form or energy product are commonly referred to as "feedstocks."

Biomass Feedstocks

Biomass feedstocks include dedicated energy crops, agricultural crop residues, forestry residues, algae, wood processing residues, municipal waste, and wet waste (crop wastes, forest residues, purpose-grown grasses, woody energy crops, algae, industrial wastes, sorted municipal solid waste (MSW), urban wood waste, and food waste).

Dedicated Energy Crops

Dedicated energy crops are non-food crops that can be grown on marginal land (land not suitable for traditional crops like corn and soybeans) specifically to provide biomass. These break down into two general categories: herbaceous and woody. Herbaceous energy crops are perennial (plants that live for more than 2 years) grasses that are harvested annually after taking 2 to 3 years to reach full productivity. These include switchgrass, miscanthus, bamboo, sweet sorghum, tall fescue, kochia, wheatgrass, and others. Short-rotation woody crops are fast-growing hardwood trees that are harvested within 5 to 8 years of planting. These include hybrid poplar, hybrid willow, silver maple, eastern cottonwood, green ash, black walnut, sweetgum, and sycamore. Many of these species can help improve water and soil quality, improve wildlife habitat relative to annual crops, diversify sources of income, and improve overall farm productivity.

Agricultural Crop Residue

There are many opportunities to leverage agricultural resources on existing lands without interfering with the production of food, feed, fiber, or forest products. Agricultural crop residues, which include the stalks and leaves, are abundant, diverse, and widely distributed across the United States. Examples include corn stover (stalks, leaves, husks, and cobs), wheat straw, oat straw, barley straw, sorghum stubble, and rice straw. The sale of these residues to a local biorefinery also represents an opportunity for farmers to generate additional income.

Forestry Residues

Forest biomass feedstocks fall into one of two categories: forest residues left after logging timber (including limbs, tops, and culled trees and tree components that would be otherwise unmerchantable) or whole-tree biomass harvested explicitly for biomass. Dead, diseased, poorly formed, and other unmerchantable trees are often left in the woods following timber harvest. This woody debris can be collected for use in bioenergy, while leaving enough behind to provide habitat and maintain proper nutrient and hydrologic features. There are also opportunities to make use of excess biomass on millions of acres of forests. Harvesting excessive woody biomass can reduce the risk of fire and pests, as well as aid in forest restoration, productivity, vitality, and resilience. This biomass could be harvested for bioenergy without negatively impacting the health and stability of forest ecological structure and function.

Algae

Algae as feedstocks for bioenergy refers to a diverse group of highly productive organisms that include microalgae, macroalgae (seaweed), and cyanobacteria (formerly called "blue-green algae"). Many use sunlight and nutrients to create biomass, which contains key components—including lipids, proteins, and carbohydrates— that can be converted and upgraded to a variety of biofuels and products. Depending on the strain, algae can grow by using fresh, saline, or brackish water from surface water sources, groundwater, or seawater. Additionally, they can grow in water from second-use sources, such as treated industrial wastewater; municipal, agricultural, or aquaculture wastewater; or produced water generated from oil and gas drilling operations.

Wood Processing Residues

Wood processing yields byproducts and waste streams that are collectively called wood processing residues and have significant energy potential. For example, the processing of wood for products or pulp produces unused sawdust, bark, branches, and leaves/needles. These residues can then be converted into biofuels or bioproducts. Because these residues are already collected at the point of processing, they can be convenient and relatively inexpensive sources of biomass for energy.

Sorted Municipal Waste

MSW resources include mixed commercial and residential garbage, such as yard trimmings, paper and paperboard, plastics, rubber, leather, textiles, and food wastes. MSW for bioenergy also represents an opportunity to reduce residential and commercial waste by diverting significant volumes from landfills to the refinery.

Wet Waste

Wet waste feedstocks include commercial, institutional, and residential food wastes (particularly those currently disposed of in landfills); organic-rich biosolids (i.e., treated sewage sludge from municipal wastewater); manure slurries from concentrated livestock operations; organic wastes from industrial operations; and biogas (the gaseous product of the decomposition of organic matter in the absence of oxygen) derived from any of the above feedstock streams. Transforming these "waste streams" into energy can help create additional revenue for rural economies and solve waste-disposal problems.

RENEWABLE ENERGY

Renewable energy uses energy sources that are continually replenished by nature—the sun, the wind, water, the Earth's heat, and plants. Renewable energy technologies turn these fuels into usable forms of energy—most often electricity, but also heat, chemicals, or mechanical power.

Importance of Renewable Energy

Today we primarily use fossil fuels to heat and power our homes and fuel our cars. It's convenient to use coal, oil, and natural gas for meeting our energy needs, but we have a limited supply of these fuels on the Earth. We're using them much more rapidly than they are being created. Eventually, they will run out and because of safety concerns and waste disposal problems, the United States will retire much of its nuclear capacity by 2020. In the meantime, the nation's energy needs are expected to grow by 33 percent during the next 20 years. Renewable energy can help fill the gap.

Even if we had an unlimited supply of fossil fuels, using renewable energy is better for the environment. We often call renewable energy technologies "clean" or "green" because they produce few if any pollutants. Burning fossil fuels, however, sends greenhouse gases into the atmosphere, trapping the sun's heat and contributing to global warming. Climate scientists generally agree that the Earth's average temperature has risen in the past century. If this trend continues, sea levels will rise, and scientists predict that floods, heat waves, droughts, and other extreme weather conditions could occur more often.

Other pollutants are released into the air, soil, and water when fossil fuels are burned. These pollutants take a dramatic toll on the environment—and on humans. Air pollution contributes to diseases like asthma. Acid rain from sulfur dioxide and nitrogen oxides harms plants and fish. Nitrogen oxides also contribute to smog.

Renewable energy will also help us develop energy independence and security. The United States imports more than 50 percent of its oil, up from 34 percent in 1973. Replacing some of our petroleum with fuels made from plant matter, for example, could save money and strengthen our energy security.

Renewable energy is plentiful, and the technologies are improving all the time. There are many ways to use renewable energy. Most of us already use renewable energy in our daily lives.

Hydropower

Hydropower is our most mature and largest source of renewable power, producing about 10 percent of the nation's electricity. Existing hydropower capacity is about 77,000 megawatts (MW). Hydropower plants convert the energy in flowing water into electricity. The most common form of hydropower uses a dam on a river to retain a large reservoir of water. Water is released through turbines to generate power. "Run of the river" systems, however, divert water from the river and direct it through a pipeline to a turbine.

Hydropower plants produce no air emissions but can affect water quality and wildlife habitats. Therefore, hydropower plants are now being designed and operated to minimize impacts on the river. Some of them are diverting a portion of the flow around their dams to mimic the natural flow of the river. But while this improves the wildlife's river habitat, it also reduces the power plant's output. In addition, fish ladders and other approaches, such as improved turbines, are being used to assist fish with migration and lower the number of fish killed.

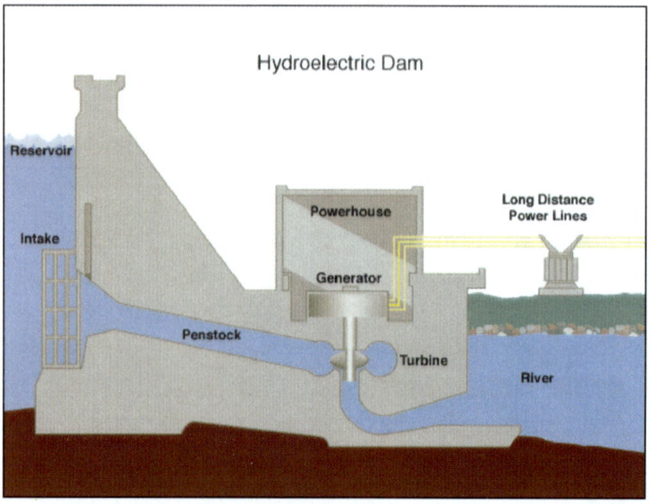

A small-scale hydropower system.

Bioenergy

Bioenergy is the energy derived from biomass (organic matter), such as plants. If you've ever burned wood in a fireplace or campfire, you've used bioenergy. But we don't get all of our biomass resources directly from trees or other plants. Many industries, such as those involved in construction or the processing of agricultural products, can create large quantities of unused or residual biomass, which can serve as a bioenergy source.

Biopower

After hydropower, biomass is this country's second-leading resource of renewable energy, accounting for more than 7,000 MW of installed capacity. Some utilities and power generating companies with coal power plants have found that replacing some coal with biomass is a low-cost option to reduce undesirable emissions. As much as 15 percent of the coal may be replaced with biomass. Biomass has less sulfur than coal. Therefore, less sulfur dioxide, which contributes to acid rain, is released into the air. Additionally, using biomass in these boilers reduces nitrous oxide emissions.

A process called gasification—the conversion of biomass into gas, which is burned in a gas turbine—is another way to generate electricity. The decay of biomass in landfills also produces gas, mostly methane, which can be burned in a boiler to produce steam for electricity generation or industrial processes. Biomass can also be heated in the absence of oxygen to chemically convert it into a type of fuel oil, called pyrolysis oil. Pyrolysis oil can be used for power generation and as a feedstock for fuels and chemical production.

Biofuels

Biomass can be converted directly into liquid fuels, called biofuels. Because biofuels are easy to transport and possess high energy density, they are favored to fuel vehicles and sometimes stationary power generation. The most common biofuel is ethanol, an alcohol made from the fermentation of biomass high in carbohydrates. The current largest source of ethanol is corn. Some cities use ethanol as a gasoline additive to help meet air quality standards for ozone.

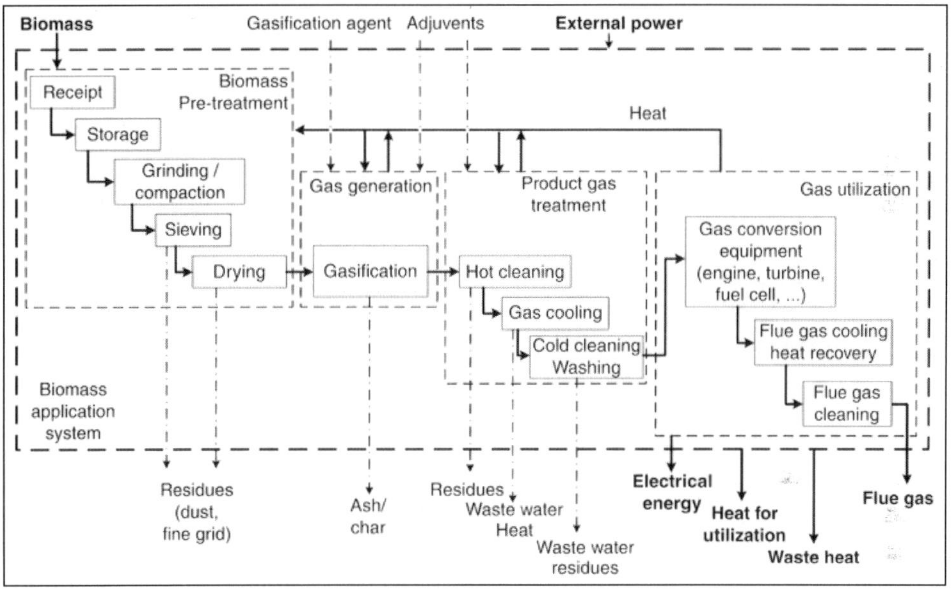

Biomass Gasification for Rural Electrification.

Flex-fuel vehicles are also now on the market, which can use a mixture of gasoline and ethanol, such as E85—a mixture of 85 percent ethanol and 15 percent gasoline. Another biofuel is biodiesel, which can be made from vegetable and animal fats. Biodiesel can be used to fuel a vehicle or as a fuel additive to reduce emissions.

Corn ethanol and biodiesel provide about 0.4 percent of the total liquid fuels market. To increase our available supply of biofuels, researchers are testing crop residues—such as cornstalks and leaves—wood chips, food waste, grass, and even trash as potential biofuel sources.

Bio-based Products

Biomass—corn, wheat, soybeans, wood, and residues—can also be used to produce chemicals and materials that we normally obtain from petroleum. Industry has already begun to use cornstarch to produce commodity plastics, such as shrinkwrap, plastic eating utensils, and even car bumpers. Commercial development is underway to make thermoset plastics, like electrical switch plate covers, from wood residues.

Geothermal Energy

The Earth's core, 4,000 miles below the surface, can reach temperatures of 9000° F. This heat—geothermal energy—flows outward from the core, heating the surrounding area, which can form underground reservoirs of hot water and steam. These reservoirs can be tapped for a variety of uses, such as to generate electricity or heat buildings. By using geothermal heat pumps (GHPs), we can even take advantage of the shallow ground's stable temperature for heating and cooling buildings.

The geothermal energy potential in the uppermost 6 miles of the Earth's crust amounts to 50,000 times the energy of all oil and gas resources in the world. In the United States, most geothermal reservoirs are located in the western states, Alaska, and Hawaii. GHPs, however, can be used almost anywhere.

Geothermal Electricity Production

Geothermal power plants access the underground steam or hot water from wells drilled a mile or more into the earth. The steam or hot water is piped up from the well to drive a conventional steam turbine, which powers an electric generator. Typically, the water is then returned to the ground to recharge the reservoir and complete the renewable energy cycle.

There are three types of geothermal power plants: dry steam, flash steam, and binary cycle. Dry steam plants draw from reservoirs of steam, while both flash steam and binary cycle plants draw from reservoirs of hot water. Flash steam plants typically use water at temperatures greater than 360°F. Unlike both steam and flash plants, binary-cycle plants transfer heat from the water to what's called a working fluid. Therefore binary cycle plants can operate using water at lower temperatures of about 225° to 360 °F.

Geothermal Power Plant.

All of the U.S. geothermal power plants are in California, Nevada, Utah, and Hawaii. Altogether about 2800 MW of geothermal electric capacity is produced annually in this country.

Geothermal Direct-use

If you've ever soaked in a natural hot spring, you're one of millions of people around the world who has enjoyed the direct use of geothermal energy. Direct-use applications require geothermal temperatures between about 70° to 302°F—lower than those required for electricity generation. The United States already has about 1,300 geothermal direct-use systems in operation.

In a direct-use system, a well is drilled into a geothermal reservoir, which provides a steady stream of hot water. Some systems use the water directly, but most pump the water through what's called a heat exchanger. The heat exchanger keeps the water separate from a working fluid (usually water or a mixture of water and antifreeze), which is heated by the geothermal water. The working fluid then flows through piping, distributing the heat directly for its intended use.

The heated water or fluid can be used in a building to replace the traditional heat source—often natural gas—of a boiler, furnace, and hot water heater. Some cities and towns actually have large direct-use heating systems—called district heating— that provide many buildings with heat. Geothermal direct use is also used in agriculture—such as for fish farms and to heat greenhouses—and for industrial food processing (vegetable dehydration).

Geothermal Heat Pumps

While air temperatures can vary widely through the seasons, the temperatures of the shallow ground only range from 50° to 70°F depending on latitude. GHPs draw on this relatively stable temperature as a source for heating buildings in the winter and keeping them cool in the summer.

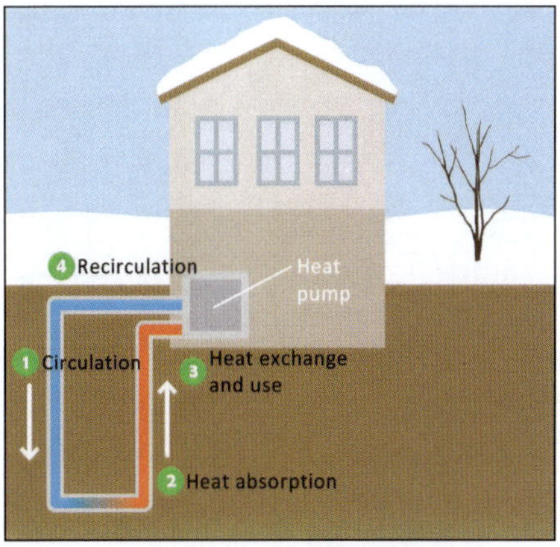

Ground source heat pump.

Through underground piping, a GHP discharges heat from inside a building into the ground in the summer, much like a refrigerator uses electricity to keep its interior cool while releasing heat into your kitchen. In the winter, this process is reversed; the GHP extracts heat from the ground and releases it into a building.

Because GHPs actually move heat between homes and the earth, instead of burning fuels, they operate very cleanly and efficiently. In fact, GHPs are at least three times more efficient than even the most energy efficient furnaces on the market today.

Solar Energy

Solar technologies tap directly into the infinite power of the sun and use that energy to produce heat, light, and power.

Passive Solar Lighting and Heating

People have used the sun to heat and light their homes for centuries. Ancient Native Americans built their dwellings directly into south-facing cliff walls because they knew the sun travels low across the southern sky in the Northern Hemisphere during the winter. They also knew the massive rock of the cliff would absorb heat in winter and protect against wind and snow. At the same time, the cliffdwelling design blocked sunlight during the summer, when the sun is higher in the sky, keeping their dwellings cool.

The modern version of this sun-welcoming design is called passive solar because no pumps, fans, or other mechanical devices are used. Its most basic features include large, south-facing windows that fill the home with natural sunlight, and dark tile or brick floors that store the sun's heat and release it back into the home at night. In the summer, when the sun is higher in the sky, window overhangs block direct sunlight, which keeps the house cool. Tile and brick floors also remain cool during the summer.

Passive solar design combined with energy efficiency will go even further. Energy-efficient features such as energysaving windows and appliances, along with good insulation and weatherstripping, can make a huge difference in energy and cost savings.

Solar Water Heating

Solar energy can be used to heat water for your home or your swimming pool. Most solar water-heating systems consist of a solar collector and a water storage tank.

Solar water-heating systems use collectors, generally mounted on a south-facing roof, to heat either water or a heat-transfer fluid, such as a nontoxic antifreeze. The heated water is then stored in a water tank similar to one used in a conventional gas or electric water-heating system.

There are basically three types of solar collectors for heating water: flatplate, evacuatedtube, and concentrating. The most common type, a flat-plate collector, is an insulated, weatherproof box containing a dark absorber plate under a transparent cover. Evacuated-tube collectors are made up of rows of parallel, transparent glass tubes. Each tube consists of a glass outer tube and an inner tube, or absorber, covered with a coating that absorbs solar energy but inhibits heat loss. Concentrating collectors for residential applications are usually parabolic-shaped mirrors (like a trough) that concentrate the sun's energy on an absorber tube called a receiver that runs along the axis of the mirrored trough and contains a heat-transfer fluid.

The Four Times Square Building in New York City usesthin-film PV panels to reduce the building's power load from the utility grid.

All three types of collectors heat water by circulating household water or a heat-transfer fluid such as a nontoxic antifreeze from the collector to the water storage tanks. Collectors do this either passively or actively.

Passive solar water-heating systems use natural convection or household water pressure to circulate water through a solar collector to a storage tank. They have no electric components that could break, a feature that generally makes them more reliable, easier to maintain, and possibly longer lasting than active systems.

An active system uses an electric pump to circulate water or nontoxic antifreeze through the system. Active systems are usually more expensive than passive systems, but they are also more efficient. Active systems also can be easier to retrofit than passive systems because their storage tanks do not need to be installed above or close to the collectors. Also, the moving water in the system will not freeze in cold climates. But because these systems use electricity, they will not function in a power outage. That's why many active systems are now combined with a small solar-electric panel to power the pump.

The amount of hot water a solar water heater produces depends on the type and size of the system, the amount of sun available at the site, proper installation, and the tilt angle and orientation of the collectors. But if you're currently using an electric water heater, solar water heating is a cost-effective alternative. If you own a swimming pool, heating the water with solar collectors can also save you money.

Solar Electricity

Solar electricity or photovoltaic (PV) technology converts sunlight directly into electricity. Solar electricity has been a prime source of power for space vehicles since the inception of the space program. It has also been used to power small electronics and rural and agricultural applications for three decades. During the last decade, a strong solar electric market has emerged for powering urban grid-connected homes and buildings as a result of advances in solar technology along with global changes in electric industry restructuring.

Although many types of solar electric systems are available today, they all consist of basically three main items: modules that convert sunlight into electricity; inverters that convert that electricity into alternating current so it can be used by most household appliances; and possibly or sometimes batteries that store excess electricity produced by the system. The remainder of the system comprises equipment such as wiring, circuit breakers, and support structures.

Today's modules can be built into glass skylights and walls. Some modules resemble traditional roof shingles, but they generate electricity, and some come with built-in inverters. The solar modules available today are more efficient and versatile than ever before.

In over 30 states, any additional power produced by a PV system, which is not being used by a home or building, can be fed back to the electric grid through a process known as net metering. Net metering allows electricity customers to pay only for their "net" electricity, or the amount of power consumed from their utility minus the power generated by their PV system. This metering arrangement allows consumers to realize full retail value for 100 percent of the PV energy produced by their systems.

Grid-connected PV systems do not require batteries. However, some grid connected systems use them for emergency backup power. And of course in remote areas, solar electricity is often a economic alternative to expensive distribution line extensions incurred by a customer first connecting to the utility grid. Electricity produced by solar electric systems in remote locations is stored in batteries. Batteries will usually store electricity produced by a solarelectric system for up to three days.

This dish/Stirling solar power system in Arizona is capable of producing 25 kW of electricity.

What type of system to purchase will depend on the energy-efficiency of your home, your home's location, and your budget. Before you size your system, try reducing energy demand through energyefficient measures. Purchasing energy-saving appliances and lights, for example, will reduce your electrical demand and allow you to purchase a smaller solar-electric system to meet your energy needs or get more value from a larger system. Energy efficiency allows you to start small and then add on as your energy needs increase.

Solar Thermal Electricity

Unlike solar-electric systems that convert sunlight into electricity, solar thermal electric systems convert the sun's heat into electricity. This technology is used primarily in large-scale power plants for powering cities and communities, especially in the Southwest where consistent hours of sunlight are greater than other parts of the United States.

Concentrating solar power (CSP) technologies convert solar energy into electricity by using mirrors to focus sunlight onto a component called a receiver. The receiver transfers the heat to a conventional engine-generator—such as a steam turbine—that generates electricity.

There are three types of CSP systems: power towers (central receivers), parabolic troughs, and dish/engine systems. A power tower system uses a large field of mirrors to concentrate sunlight onto the top of a tower, where a receiver sits. Molten salt flowing through the receiver is heated by the concentrated sunlight. The salt's heat is turned into electricity by a conventional steam generator. Parabolictrough systems concentrate the sun's energy through long, parabolic-shaped mirrors. Sunlight is focused on a pipe filled with oil that runs down the axis of the trough. When the oil gets hot, it is used to boil water in a conventional steam generator to produce electricity. A dish/engine system uses a mirrored dish (similar in size to a large satellite dish). The dish-shaped surface focuses and concentrates the sun's heat onto a receiver at the focal point of the dish (above and center of the collectors). The receiver absorbs the sun's heat and transfers it to a fluid within an engine, where the heat causes the fluid to expand against a piston to produce mechanical power. The mechanical power is then used to run a generator or alternator to produce electricity.

Concentrating solar technologies can be used to generate electricity for a variety of applications, ranging from remote power systems as small as a few kilowatts (kW) up to grid-connected applications of 200 MW or more. A 354-MW power plant in Southern California, which consists of nine trough power plants, meets the energy needs of more than 350,000 people and is the world's largest solar energy power plant.

Wind Energy

For hundreds of years, people have used windmills to harness the wind's energy. Today's wind turbines, which operate differently from windmills, are a much more efficient technology.

Wind turbine technology may look simple: the wind spins turbine blades around a central hub; the hub is connected to a shaft, which powers a generator to make electricity. However, turbines are highly sophisticated power systems that capture the wind's energy by means of new blade designs or airfoils. Modern, mechanical drive systems, combined with advanced generators, convert that energy into electricity.

The 6-MW Green Mountain power plant in Searsburg,Vermont, consists of eleven 550-kW wind turbines.

Wind turbines that provide electricity to the utility grid range in size from 50 kW to 1 or 2 MW. Large, utility-scale projects can have hundreds of turbines spread over many acres of land. Small turbines, below 50 kW, are used to charge batteries, electrify homes, pump water for farms and ranches, and power remote telecommunications equipment. Wind turbines can also be placed in the shallow water near a coastline if open land is limited, such as in Europe, and/or to take advantage of strong, offshore winds.

Wind energy has been the fastest growing source of energy in the world since 1990, increasing at an average rate of over 25 percent per year. It's a trend driven largely by dramatic improvements in wind technology. Currently, wind energy capacity amounts to about 2500 MW in the United States. Good wind areas, which cover 6 percent of the contiguous U.S. land area, could supply more than one and a half times the 1993 electricity consumption of the entire country.

California now has the largest number of installed turbines. Many turbines are also being installed across the Great Plains, reaching from Montana east to Minnesota and south through Texas, to take advantage of its vast wind resource. North Dakota alone has enough wind to supply 36 percent of the total 1990 electricity consumption of the lower 48 states. Hawaii, Iowa, Minnesota, Oregon, Texas, Washington, Wisconsin, and Wyoming are among states where wind energy use is rapidly increasing.

Hydrogen

Hydrogen is high in energy, yet its use as a fuel produces water as the only emission. Hydrogen is the universe's most abundant element and also its simplest. A hydrogen atom consists of only one proton and one electron. Despite its abundance and simplicity, it doesn't occur naturally as a gas on the Earth.

Today, industry produces more than 4 trillion cubic feet of hydrogen annually. Most of this hydrogen is produced through a process called reforming, which involves the application of heat to separate hydrogen from carbon. Researchers are developing highly efficient, advanced reformers to produce hydrogen from natural gas for what's called Proton Exchange Membrane fuel cells.

You can think of fuel cells as batteries that never lose their charge. Today, hydrogen fuel cells offer tremendous potential to produce electrical power for distributed energy systems and vehicles. In the future, hydrogen could join electricity as an important "energy carrier": storing, moving, and delivering energy in a usable form to consumers. Renewable energy sources, like the sun, can't produce energy all the time. But hydrogen can store the renewable energy produced until it's needed.

Eventually, researchers would like to directly produce hydrogen from water using solar, wind, and biomass and biological technologies.

Ocean Energy

The ocean can produce two types of energy: thermal energy from the sun's heat, and mechanical energy from the tides and waves.

Ocean thermal energy can be used for many applications, including electricity generation. Electricity conversion systems use either the warm surface water or boil the seawater to turn a turbine, which activates a generator.

The electricity conversion of both tidal and wave energy usually involves mechanical devices. A dam is typically used to convert tidal energy into electricity by forcing the water through turbines, activating a generator. Meanwhile, wave energy uses mechanical power to directly activate a generator, or to transfer to a working fluid, water, or air, which then drives a turbine/generator.

Most of the research and development in ocean energy is happening in Europe.

BIOENERGY

Bioenergy is renewable energy created from natural, biological sources. Many natural sources, such as plants, animals, and their byproducts, can be valuable resources. Modern technology even makes landfills or waste zones potential bioenergy resources. It can be used to be a sustainable power source, providing heat, gas, and fuel. Because the energy contained in sources like plants is obtained from the sun through photosynthesis, it can be replenished and is considered an inexhaustible source.

Using bioenergy has the potential to decrease our carbon footprint and improve the environment. While bioenergy uses the same amount of carbon dioxide as traditional fossil fuels, the impact can be minimized as long as the plants used are replaced. Fast-growing trees and grass help this process and are known as bioenergy feedstocks.

Most bioenergy comes from forests, agricultural farms, and waste. The feedstocks are grown by farms specifically for their use as an energy source. Common crops include starch or sugar-based plants, like sugarcane or corn.

To turn raw sources into energy, there are three processes: chemical, thermal, and biochemical. Chemical processing uses chemical agents to break down the natural source and convert it into liquid fuel. Corn ethanol, a fuel created from corn, is an example of chemical processing results. Thermal conversion uses heat to change the source into energy through combustion or gasification. Biochemical conversion uses bacteria or other organisms to convert the source, such as through composting or fermentation.

Bioenergy exists at several different levels. Individuals can create bioenergy, such as by creating a compost heap out of kitchen scraps and keeping worms to produce rich fertilizer. At the other extreme are large energy corporations looking for more sustainable energy sources than oil or coal. These organizations use huge farms and facilities to provides hundreds or thousands of customers with energy.

Importance of Bioenergy

Having the ability to produce energy through plants or other resources can lessen U.S. reliance on foreign nations for sources of energy. Bioenergy also is viewed as essential for the environment. Continued use of fossil fuels can cause significant environmental issues by producing greenhouse gases that contribute to global warming or by emitting harmful pollutants such as sulfur dioxide which can harm the population's health.

As technology progresses, bioenergy has the potential to dramatically reduce greenhouse emissions, the release of harmful gases associated with global warming and climate change. The use of forests and farms in bioenergy can help combat the harmful release of carbon dioxide and help achieve a balance.

At this time, bioenergy is not ready to replace fossil fuels. The process is too costly and uses too many resources to be practical in most areas. The large plots of land and significant amounts of water needed to be successful can be difficult for many states or countries. Additionally, agricultural resources such as land and water dedicated to producing crops related to bioenergy can limit the resources used for producing food. Still, as science continues to study this area, bioenergy could increasingly become a larger source of energy that can help improve the environment.

Abundant and renewable bioenergy can contribute to a more secure, sustainable, and economically sound future by:

- Supplying domestic clean energy sources.
- Reducing U.S. dependence on foreign oil.
- Generating U.S. jobs.
- Revitalizing rural economies.

The U.S. Department of Energy's 2016 Billion-Ton Report: Advancing Domestic Resources for a Thriving Bioeconomy concluded that the United States has the potential to produce 1 billion dry tons of non-food biomass resources annually by 2040 and still meet demands for food, feed, and fiber. One billion tons of biomass could:

- Produce up to 50 billion gallons of biofuels.
- Yield 50 billion pounds of bio-based chemicals and bioproducts.
- Generate 85 billion kilowatt-hours of electricity to power 7 million households.
- Contribute 1.1 million jobs to the U.S. economy.
- Keep $260 billion in the United States.

Biofuels: Energy for Transportation

Biomass is one type of renewable resource that can be converted into liquid fuels—known as biofuels—for transportation. Biofuels include cellulosic ethanol, biodiesel, and renewable hydrocarbon "drop-in" fuels. The two most common types of biofuels in use today are ethanol and biodiesel. Biofuels can be used in airplanes and most vehicles that are on the road. Renewable transportation fuels that are functionally equivalent to petroleum fuels lower the carbon intensity of our vehicles and airplanes.

Biopower: Energy for Heat and Electricity

Biopower technologies convert renewable biomass fuels into heat and electricity using processes like those used with fossil fuels. There are three ways to harvest the energy stored in biomass to produce biopower: burning, bacterial decay, and conversion to a gas or liquid fuel. Biopower can offset the need for carbon fuels burned in power plants, thus lowering the carbon intensity of electricity generation. Unlike some forms of intermittent renewable energy, biopower can increase the flexibility of electricity generation and enhance the reliability of the electric grid.

Bioproducts: Everyday Commodities made from Biomass

Biomass is a versatile energy resource, much like petroleum. Beyond converting biomass to biofuels for vehicle use, it can also serve as a renewable alternative to fossil fuels in the manufacturing of bioproducts such as plastics, lubricants, industrial chemicals, and many other products currently derived from petroleum or natural gas. Mimicking the existing petroleum refinery model, integrated biorefineries can produce bioproducts alongside biofuels. This co-production strategy offers a more efficient, cost-effective, and integrated approach to the use of U.S. biomass resources. Revenue generated from bioproducts also offers added value, improving the economics of biorefinery operations and creating more cost-competitive biofuels.

BIOMASS COGENERATION

Cogeneration refers to the process wherein we obtain both heat and electricity from the same fuel at the same time. The process is also referred to as CHP, short for combined heat and power. A variety of fuels can be used for cogeneration including bagasse, natural gas, coal, and biomass. A cogeneration plant consists of four basic elements: a prime mover, an electricity generator, a heat extraction or recovery unit, and a control panel.

Fuel is burnt in the system or prime mover to convert its chemical energy into heat energy, which, in turn, produces the mechanical energy to run a generator and ultimately produce electricity. Prime movers for CHP systems include steam turbines, gas turbines, reciprocating engines, micro-turbines, and fuel cells. The heat energy from the system is also used directly, as heat, or indirectly to produce steam, hot water, and hot air, thus making it a CHP or cogeneration system.

Cogeneration Technologies

Different types of cogeneration technologies are used depending upon the end use or purpose. Some commonly used cogeneration technologies are:

- Steam Turbines
- Gas Turbines
- Reciprocating Engine

Steam turbine cogeneration is the commonly used technology. However, the other two technologies are also discussed for information.

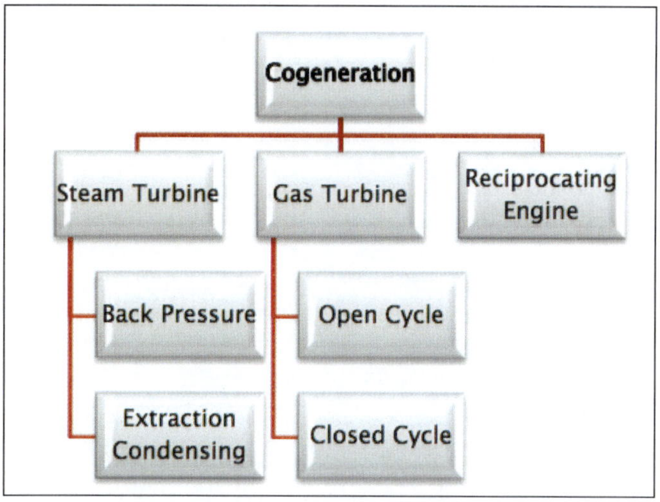

Steam Turbine Cogeneration Systems

Steam turbines work on the principle of the Rankine cycle, which consists of a heat source (boiler) that converts water into high-pressure steam. A multistage turbine allows the high-pressure steam to expand, which lowers its pressure. The steam is then transported to a condenser, which is like a vacuum chamber and thus has negative pressure and converts, or condenses, the steam into water. Alternatively, the steam may be transported to a distribution system that delivers steam at intermediate temperatures for different applications. The condensate from the condenser or from the steam utilization system returns to the feed water pump, and the cycle continues. These systems are suitable for capacities of 500 kW to 100 MW or even higher.

Most common steam turbines used in a cogeneration system are the back-pressure type or the extraction–condensing type. The choice between the two types depends on how much electricity and heat are required, steam pressure and temperature, and the economics of operation.

Back-pressure Steam Turbines

Steam at a pressure higher or equal to atmospheric pressure is extracted from the turbine to the thermal load that is the point at which heat is required. At that point, the steam releases heat and gets condensed, or turns into water. The condensate (water) returns to the system at a flow rate that can be lower than the steam flow rate if some steam is used in the process. This loss of steam is then compensated for in the cycle in the form of 'make-up' water fed into the boiler. It has to be noted that this turbine system does not have a separate condenser. Back-pressure steam turbines are the most efficient among all cogeneration systems; their cogeneration efficiency ranges from 84% to 92%.

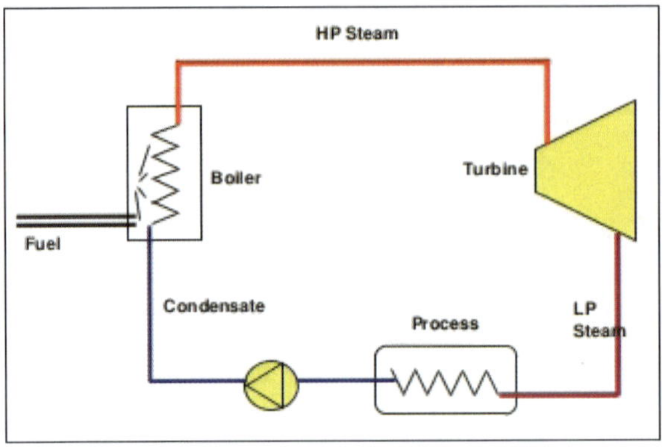

Back–pressure steam turbine systems.

Extraction-condensing Steam Turbines

In extraction-condensing steam turbines, steam is extracted at one or more intermediate stages at the required pressure and temperature. The remaining steam from the turbine is transported to the condenser at very low pressure, as low as 0.05 bar (5 kPa), corresponding to a condensing temperature of approximately 33 °C.

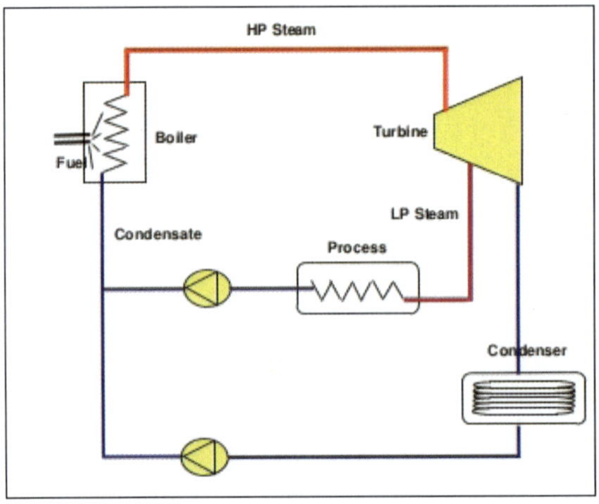

Extraction-condensing steam turbine.

Gas Turbine Cogeneration Systems

Gas turbine cogeneration systems work on the principle of the Brayton cycle, in which atmospheric air is compressed, heated, and then expanded, producing more power than what is consumed by the compressor in compressing and heating the air. The capacity of gas turbines varies from a fraction of a megawatt to about 100 MW. A variety of fuels can be used: natural gas, light petroleum distillates such as gas oil and diesel oil, products of coal gasification, etc. Gas turbine cogeneration systems are often more useful than steam turbines because gas turbine systems are more flexible; they can operate at widely varying ratios of electrical output to thermal output as required by the intended use.

Open Cycle Gas Turbine Cogeneration Systems

Most of the currently available gas turbine systems work on the open Brayton cycle, in which the compressor takes in air from the atmosphere and sends the compressed air to the combustor. The air temperature also increases because of compression. Older and smaller units operate at a pressure ratio (ratio of outlet air pressure to inlet air pressure) of 15:1, whereas the newer and larger units operate at pressure ratios approaching 30:1.

From the compressor, the air is delivered through a diffuser to a combustion chamber, where fuel is injected and burnt. Exhaust gases exit the combustor at high temperatures (about 600 °C). The highest temperature in the cycle is reached at this point; the higher this temperature, the higher the cycle's efficiency. This temperature is limited – currently to about 1300 °C – by the ability of the gas turbine material to withstand high temperatures and the efficiency of the cooling blades.

The exhaust gases at high pressure and temperature enter the gas turbine, supplying the mechanical energy to drive the compressor and the electric generator, which, in turn, produces electricity. The exhaust gases leave the turbine at considerably high temperatures (450–600 °C), which are ideal for high-temperature heat recovery. The heat is recovered by a heat-recovery boiler for extracting heat more efficiently.

The steam produced in the heat recovery boiler can be at high pressures and temperatures, which makes the steam suitable not only for thermal processes but also for running a steam turbine to produce additional power. The exhaust gases are finally released into the atmosphere, after extracting maximum heat in the various components of the cogeneration system.

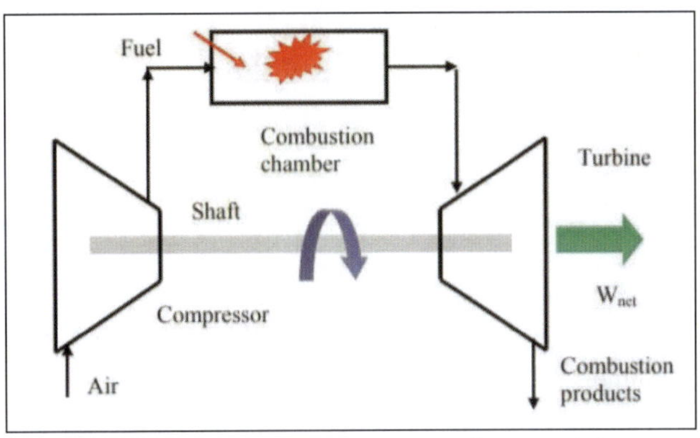

Cogeneration system with open cycle gas turbine.

Closed Cycle Gas Turbine Cogeneration Systems

In the closed cycle system, the working fluid (usually helium or air) circulates within a closed circuit. The heat is supplied to the closed cycle through a heat exchanger, instead of direct combustion of the fuel in the working fluid circuit. This arrangement ensures that both the working fluid and the turbine machinery are isolated from both the combustion chambers (heat source). On exiting from the turbine, the working fluid cools down, releasing its useful heat in the form of mechanical energy to produce electricity. In the closed cycle gas turbine, the gas turbine exhaust is recycled to the compressor after being cooled and thereby forms a closed working fluid circuit. The source of heat can be the external combustion of any fuel (e.g. industrial wastes, municipal waste, solar energy or nuclear energy). The capacities of such systems range from 2 to 50 MWe.

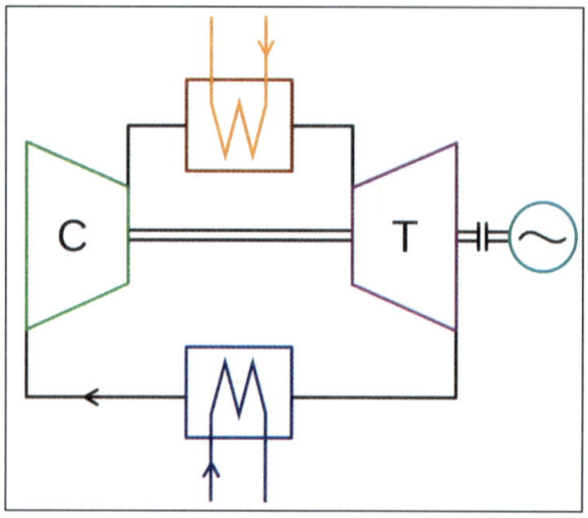

Cogeneration system with closed cycle gas turbine.

Reciprocating Engine Cogeneration Systems

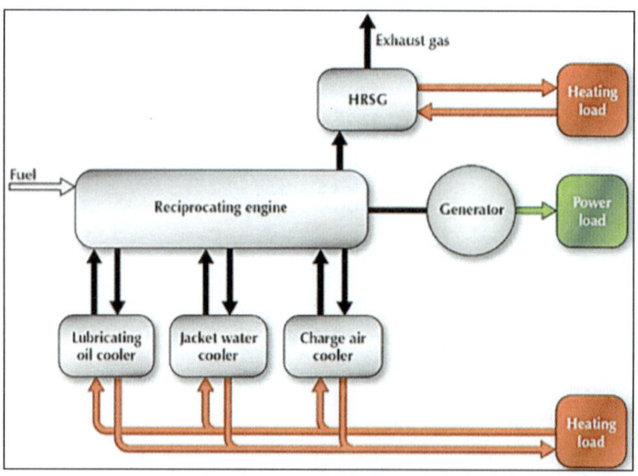

Reciprocating engine cogeneration systems.

A reciprocating engine, such as a diesel engine, can be combined with a heat-recovery boiler that supplies heat to the steam turbine to generate both electricity and heat. Heat from reciprocating engines can be recovered from four potential sources: exhaust gases, water from the engine jacket used for cooling, lube oil used for cooling, and the turbocharger used for cooling. The first two are the major sources, which are also easy to use and hence more common. Of the total heat lost from an engine (depending on its operating efficiency), roughly half is in the form of exhaust gases (400–500 °C), which can be utilized for producing steam or for drying bricks, ceramics, animal feed, etc. The waste heat in the form of the water used for cooling the engine (20%–30%) can be utilized for pre-heating water or generating hot air. Heat from this source can be used for some industrial processes that require low-pressure steam, in hospitals for sterilizing surgical equipment, garments, etc., and in food processing. Reciprocating engine cogeneration plants can attain overall efficiencies of more than 80%–90%, and their capacities span a wide range, from as little as a few kW to MW, depending on the capacity of the internal combustion engine.

Classification of Cogeneration Systems

Based on priority in utilizing the available energy, electricity or heat, cogeneration systems are classified as topping cycle (where priority is for generating electricity) and bottoming cycle (where heat takes priority over electricity).

Topping Cycle

The topping cycle is the most commonly used method of cogeneration. In this cycle, fuel is used first for producing electricity and then for heat. Steam turbine topping cycles are commonly used in the pulp and paper industry; heat recovery and combined cycle systems are used in many chemical plants; and gas turbine cycles are useful in central heating or cooling systems.

Bottoming Cycle

In a bottoming cycle, fuel is first used to produce thermal energy, and the heat rejected from the process is used for generating power through a heat-recovery boiler and a turbine generator. Bottoming cycles are suitable for manufacturing processes in which heat is rejected in large amounts and at high temperatures, typically in cement, steel, ceramic, gas, and petrochemical industries. Bottoming cycle plants are much less common than topping cycle plants.

Applications

- Industrial applications of cogeneration are found mainly in sugar factories, foodprocessing plants, pharmaceuticals, oil refineries, textile mills, and steel, cement, glass, and ceramics plants, which require both heat and electricity in substantial amounts.
- Residential, commercial, and institutional applications tend to be found in smaller systems, often based on 'packaged' units. These systems are commonly used in hotels, leisure centres, offices, smaller hospitals, and residential complexes. Larger applications are based on a technology similar to the cogeneration systems used in industry, gas turbines, or larger reciprocating engines. Such systems are used in large hospitals, large office complexes, universities, and colleges.

- District heating systems are used at airports, office and commercial buildings, and large housing complexes. The heat provided by cogeneration is ideal for space heating and for providing hot water for domestic, commercial, or industrial use. A feature of cogeneration-driven district heating systems is the option to use a variety of fuels to suit environmental, economic, or strategic priorities.

Advantages and Disadvantages of Various Cogeneration Systems

The advantages and disadvantages of different types of cogeneration systems are summarized below:

Type of cogeneration system	Advantages	Disadvantages
Back-pressure steam turbine	• Simple configuration, • Low capital cost, • High cogeneration efficiency.	• Low electrical efficiency, • Low part-load performance, • Limited flexibility in design and operation.
Extraction-condensing steam turbine	• Flexible in design and operation.	• Cost intensive.
Gas turbine	• Good fuel efficiency, • Relatively low investment cost per unit of electric output, • Short gestation period, • Low emissions, • High flexibility in operation.	• High fuel cost, • Poor efficiency at low loads, • Longer operation, • High maintenance cost.
Reciprocating engine cogeneration system	• Low civil construction cost due to block foundations and least number of auxiliaries, • High power efficiency, • Better suited as a standby power source.	• Low overall efficiency, • Availability of low temperature steam, • Highly maintenance prone.

References

- Biomass-home, energyexplained: eia.gov, Retrieved 1 May, 2019
- Dale, b.e. (2008). "grassoline in your tank: myths and realities about biofuels". Microscopy and microanalysis: 1484–1485. Doi:10.1017/s1431927608088764
- Biomass-electricity-generation, resources: wbdg.org, Retrieved 4 February, 2019
- Advantages-disadvantages-biomassenergy: conserve-energy-future.com, Retrieved 4 January, 2019
- "Mckinsey & company releases new data about growth in industrial biotech sector at world congress" retrieved 23 June 2015
- Biomass-resources, bioenergy: energy.gov, Retrieved 13 February, 2019
- What-is-bioenergy-2941107: thebalance.com, Retrieved 17 March, 2019
- Bioenergy-basics, bioenergy: energy.gov, Retrieved 20 January, 2019

Significant Aspects of Biomass

CHAPTER 2

Some of the major concepts related to biomass include bioconversion, energy forestry, bio-energy with energy capture and storage, ethanol fuel energy balance, and biomass allocation. The chapter closely examines these key concepts to provide an extensive understanding of the subject.

BIOREFINERY

A biorefinery is a facility that integrates biomass conversion processes and equipment to produce fuels, power, and value-added chemicals from biomass. Biorefinery is analogous to today's petroleum refinery, which produces multiple fuels and products from petroleum. By producing several products, a biorefinery takes advantage of the various components in biomass and their intermediates, therefore maximizing the value derived from the biomass feedstock.

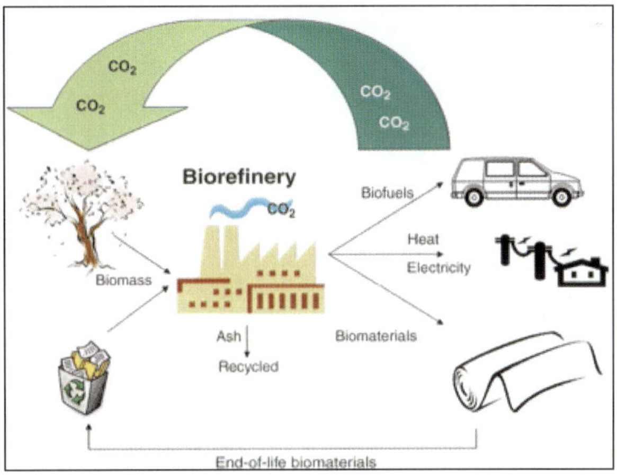

A biorefinery could, for example, produce one or several low-volume, but high-value, chemical products and a low-value, but high-volume liquid transportation fuel such as biodiesel or bioethanol. At the same time, it can generate electricity and process heat, through CHP technology, for its own use and perhaps enough for sale of electricity to the local utility. The high value products increase profitability, the high-volume fuel helps meet energy needs, and the power production helps to lower energy costs and reduce GHG emissions from traditional power plant facilities.

Biorefinery Platforms

There are several platforms which can be employed in a biorefinery with the major ones being the sugar platform and the thermochemical platform (also known as syngas platform).

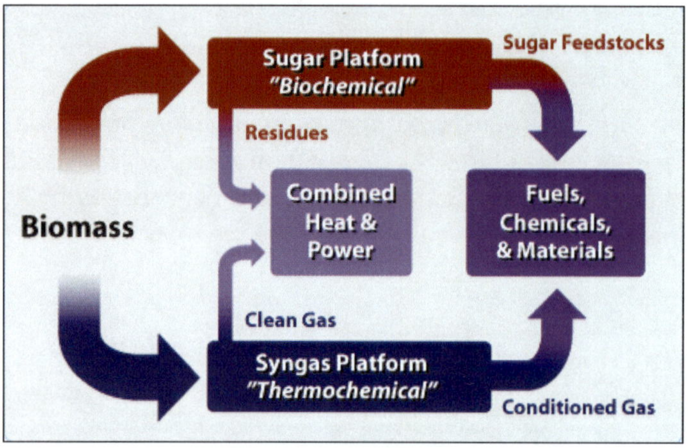

Biorefinery Platforms.

Sugar platform biorefineries breaks down biomass into different types of component sugars for fermentation or other biological processing into various fuels and chemicals. On the other hand, thermochemical biorefineries transform biomass into synthesis gas (hydrogen and carbon monoxide) or pyrolysis oil.

The thermochemical biomass conversion process is complex, and uses components, configurations, and operating conditions that are more typical of petroleum refining. Biomass is converted into syngas, and syngas is converted into an ethanol-rich mixture. However, syngas created from biomass contains contaminants such as tar and sulphur that interfere with the conversion of the syngas into products. These contaminants can be removed by tar-reforming catalysts and catalytic reforming processes. This not only cleans the syngas, it also creates more of it, improving process economics and ultimately cutting the cost of the resulting ethanol.

Advantages

Biorefineries can help in utilizing the optimum energy potential of organic wastes and may also resolve the problems of waste management and GHGs emissions. Biomass wastes can be converted, through appropriate enzymatic/chemical treatment, into either gaseous or liquid fuels. The pre-treatment processes involved in biorefining generate products like paper-pulp, HFCS, solvents, acetate, resins, laminates, adhesives, flavour chemicals, activated carbon, fuel enhancers, undigested sugars etc. which generally remain untapped in the traditional processes. The suitability of this process is further enhanced from the fact that it can utilize a variety of biomass resources, whether plant-derived or animal-derived.

BIOREACTOR

A bioreactor refers to any manufactured device or system that supports a biologically active environment. In one case, a bioreactor is a vessel in which a chemical process is carried out which involves organisms or biochemically active substances derived from such organisms. This process can either be aerobic or anaerobic. These bioreactors are commonly cylindrical, ranging in size from litres to cubic metres, and are often made of stainless steel.

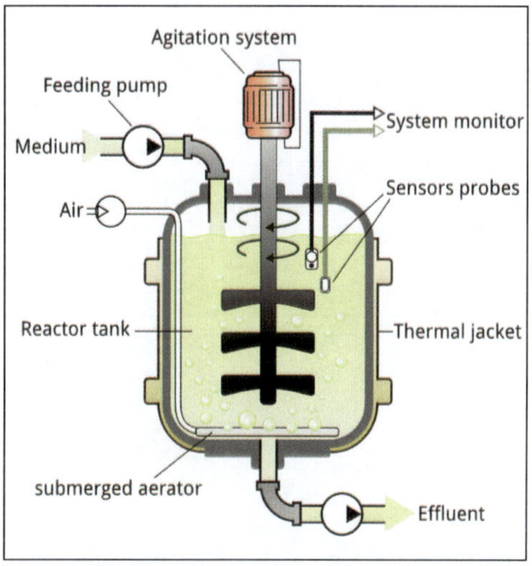

General structure of a continuous stirred-tank type bioreactor.

It may also refer to a device or system designed to grow cells or tissues in the context of cell culture. These devices are being developed for use in tissue engineering or biochemical engineering.

On the basis of mode of operation, a bioreactor may be classified as batch, fed batch or continuous (e.g. a continuous stirred-tank reactor model). An example of a continuous bioreactor is the chemostat.

Organisms growing in bioreactors may be submerged in liquid medium or may be attached to the surface of a solid medium. Submerged cultures may be suspended or immobilized. Suspension bioreactors can use a wider variety of organisms, since special attachment surfaces are not needed, and can operate at much larger scale than immobilized cultures. However, in a continuously operated process the organisms will be removed from the reactor with the effluent. Immobilization is a general term describing a wide variety of cell or particle attachment or entrapment. It can be applied to basically all types of biocatalysis including enzymes, cellular organelles, animal and plant cells. Immobilization is useful for continuously operated processes, since the organisms will not be removed with the reactor effluent, but is limited in scale because the microbes are only present on the surfaces of the vessel.

Large scale immobilized cell bioreactors are:

- Moving media, also known as moving bed biofilm reactor (MBBR),
- Packed bed,
- Fibrous bed,
- Membrane.

Design

A closed bioreactor used in cellulosic ethanol research.

Bioreactor design is a relatively complex engineering task, which is studied in the discipline of biochemical engineering. Under optimum conditions, the microorganisms or cells are able to perform their desired function with limited production of impurities. The environmental conditions inside the bioreactor, such as temperature, nutrient concentrations, pH, and dissolved gases (especially oxygen for aerobic fermentations) affect the growth and productivity of the organisms. The temperature of the fermentation medium is maintained by a cooling jacket, coils, or both. Particularly exothermic fermentations may require the use of external heat exchangers. Nutrients may be continuously added to the fermenter, as in a fed-batch system, or may be charged into the reactor at the beginning of fermentation. The pH of the medium is measured and adjusted with small amounts of acid or base, depending upon the fermentation. For aerobic (and some anaerobic) fermentations, reactant gases (especially oxygen) must be added to the fermentation. Since oxygen is relatively insoluble in water (the basis of nearly all fermentation media), air (or purified oxygen) must be added continuously. The action of the rising bubbles helps mix the fermentation medium and also "strips" out waste gases, such as carbon dioxide. In practice, bioreactors are often pressurized; this increases the solubility of oxygen in water. In an aerobic process, optimal oxygen transfer is sometimes the rate limiting step. Oxygen is poorly soluble in water—even less in warm fermentation broths—and is relatively scarce in air (20.95%). Oxygen transfer is usually helped by agitation, which is also needed to mix nutrients and to keep the fermentation homogeneous. Gas dispersing agitators are used to break up air bubbles and circulate them throughout the vessel.

Fouling can harm the overall efficiency of the bioreactor, especially the heat exchangers. To avoid it, the bioreactor must be easily cleaned. Interior surfaces are typically made of stainless steel for easy cleaning and sanitation. Typically bioreactors are cleaned between batches, or are designed to reduce fouling as much as possible when operated continuously. Heat transfer is an important part of bioreactor design; small vessels can be cooled with a cooling jacket, but larger vessels may require coils or an external heat exchanger.

Modeling

Mathematical models act as an important tool in various bio-reactor applications including wastewater treatment. These models are useful for planning efficient process control strategies and predicting the future plant performance. Moreover, these models are beneficial in education and research areas.

Bioreactors are generally used in those industries which are concerned with food, beverages and pharmaceuticals. The emergence of *Biochemical engineering* is of recent origin. Processing of biological materials using biological agents such as cells, enzymes or antibodies are the major pillars of biochemical engineering. Applications of biochemical engineering cover major fields of civilization such as agriculture, food and healthcare, resource recovery and fine chemicals.

Till now, the industries associated with biotechnology have been lagged behind other industries in implementing control over the process and optimization strategies. A main drawback in biotechnological process control is the problem to measure key physical and biochemical parameters.

Operational Stages in a Bio-process

A bioprocess is composed mainly of three stages — upstream processing, bioreaction, and downstream processing — to convert raw material to finished product. The raw material can be of biological or non-biological origin. It is first converted to more suitable form for processing. This is done in upstream processing step which involves chemical hydrolysis, preparation of liquid medium, separation of particulate, air purification and many other preparatory operations.

After upstream processing step, the resulting feed is transferred to one or more Bioreaction stages. The Biochemical reactors or bioreactors form the base of the Bioreaction step. This step is mainly consists of three operations namely, production of biomass, metabolize biosynthesis and biotransformation.

Finally, the material produced in the bioreactor must be further processed in the downstream section to convert it into more useful form. The downstream process is mainly consists of physical separation operations which includes, solid liquid separation, adsorption, liquid-liquid extraction, distillation, drying etc.

Specifications

A typical bioreactor consists of following parts:

- Agitator – Used for the mixing of the contents of the reactor which keeps the "cells" in the perfect homogenous condition for better transport of nutrients and oxygen to the desired products.
- Baffle – Used to break the vortex formation in the vessel, which is usually highly undesirable as it changes the center of gravity of the system and consumes additional power.
- Sparger – In aerobic cultivation process, the purpose of the sparger is to supply adequate oxygen to the growing cells.
- Jacket – The jacket provides the annular area for circulation of constant temperature of water which keeps the temperature of the bioreactor at a constant value.

Development of Modeling Equations for Bioreactors

Assumptions:

- The reactor contents are perfectly mixed together.
- The reactor is operating at a constant temperature (i.e., it is isothermal).
- The feed should be clean and pure (i.e., no biomass in the feed stream).
- The feed stream and reactor contents have equal and constant density (ρ).
- The feed and product streams have the same flow rate (F).

Total Continuity Equation

Making overall mass balance, we get the following equation:

$$d(\rho V)/dt = F\rho - F\rho = 0$$

Equation states that the reactor volume (V) is constant since $dV/dt = 0$.

Biomass Continuity Equation

We know,

- Flow rate of biomass into the reactor = Fx_i
- Flow rate of biomass out of the reactor = Fx
- Rate of generation of biomass by reaction = Vr_1
- Rate of accumulation of biomass within the reactor = $d(Vx)/dt$

Now, apply general mass balance equation i.e.,

Rate of Mass In – Rate of Mass Out + Rate of Generation = Accumulation

$$d(Vx)/dt = Fx_i - Fx + Vr_1$$

where r_1 is the rate of cell generation. Dividing both sides of the above equation by V, we obtain:

$$dx/dt = (F/V)x_i - (F/V)x + r_1$$

In the chemical reaction engineering, F/V is called space velocity(s^{-1}) and V/F is called the residence time (s). But in biochemical engineering, F/V is known as Dilution rate (D_r). Accordingly, equation yields:

$$dx/dt = D_r x_i - D_r x + r_1$$

$$dx/dt = D_r(x_i - x) + r_1$$

Substrate Continuity Equation

For substrate balance,

- Flow rate of substrate into the bioreactor = FS_i
- Flow of the substrate out of the bioreactor = FS
- Rate of generation of substrate by reaction = $-Vr_2$
- Rate of accumulation of substrate within the reactor = $d(VS)/dt$

Now, apply general mass balance equation i.e.,

Rate of Mass In – Rate of Mass Out + Rate of Generation = Accumulation

$$d(VS)/dt = FS_i - FS - Vr_2$$

Rearranging above equation, we get:

$$dS/dt = D_r(S_i - S) - r_2$$

where r_2 is the rate of substrate consumption.

Biochemical Reaction Kinetics

For the chemical reaction,

$$A \text{ ----> } P$$

We can write:

$$(-r_A) = k\,(C_A)^n$$

$$(r_A) = -\,k\,(C_A)^n$$

where,

- $(-r_A)$ = rate of disappearance of A
- (r_A) = rate of formation of A
- k = reaction rate constant
- C_A = Concentration of reactant A
- n = order of reaction with respect to component A

For first order reaction, n = 1 and accordingly,

$$-r_A = k\,C_A$$

The reaction kinetics involved in biochemical operations is comparatively difficult to obtain than the chemical reaction kinetics. In biochemical operations, the cell kinetics is used for the unstructured models where balanced growth condition is assumed.

The following equation is used to represent the net rate of cell mass growth:

$$r_1 = \mu x$$

where μ is the specific growth rate or specific growth rate coefficient(s^{-1}). Here, μ is analogous to first order rate constant k but however, μ is not a constant.

In biochemical engineering, yield is defined as the ratio of mass or moles of product formed to the mass or moles of the reactants consumed. The yield (Y) of product (P) with respect to reactant A is defined as:

$$Y = (\text{mass of P formed})/(\text{mass of A consumed})$$

In case of bioreactor,

$$Y = (\text{mass of cells formed})/(\text{mass of substrate consumed})$$

Thus,

$$Y = r_1 / r_2$$

Or,

$$r_2 = r_1/Y$$

Or,

$$r_2 = \mu x/Y$$

Final form of Equation of Modeling

By substituting equations in equations respectively, we get,

$$dx/dt = D_r(x_i - x) + \mu x$$

$$dS/dt = D_r(S_i - S) - (\mu x/Y)$$

Since we have assumed that the feed stream does not contain any biomass i.e., $x_i = 0$, then, bioreactor modelling equation finally get the following form:

$$dx/dt = (\mu - D_r)x$$

$$dS/dt = D_r(S_i - S) - (\mu x/Y)$$

Thus, equations $dS/dt = D_r(S_i - S) - (\mu x/Y)$ and $dx/dt = (\mu - D_r)x$ are the basic equations which are used for the modelling of any bioreactor.

Types of Bioreactors

The different types of bioreactors are:

Continuous Stirred Tank Bioreactors

A continuous stirred tank bioreactor consists of a cylindrical vessel with motor driven central shaft that supports one or more agitators (impellers). The shaft is fitted at the bottom of the bioreactor. The number of impellers is variable and depends on the size of the bioreactor i.e., height to diameter ratio, referred to as aspect ratio.

The aspect ratio of a stirred tank bioreactor is usually between 3-5. However, for animal cell culture applications, the aspect ratio is less than 2. The diameter of the impeller is usually 1/3 rd of the vessel diameter. The distance between two impellers is approximately 1.2 impeller diameter. Different types of impellers (Rustom disc, concave bladed, marine propeller etc.) are in use.

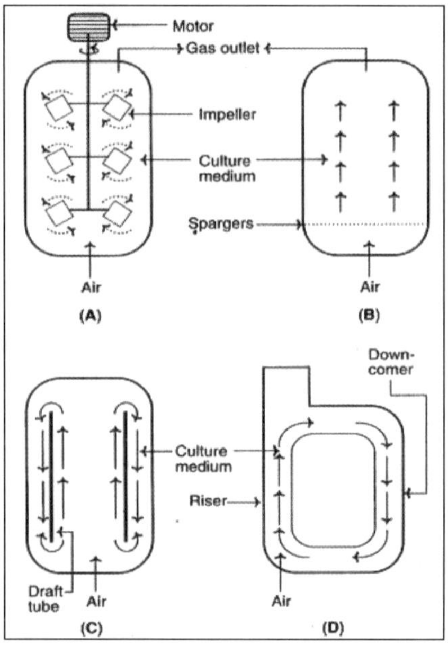

Type of bioreactors: (A) Continuous stirred tank bioreactor (B) Bubble column bioreactor (C) Internal-loop airlift bioreactor (D) External-loop airlift bioreactor.

In stirred tank bioreactors or in short stirred tank reactors (STRs), the air is added to the culture medium under pressure through a device called sparger. The sparger may be a ring with many holes or a tube with a single orifice. The sparger along with impellers (agitators) enables better gas distribution system throughout the vessel.

The bubbles generated by sparger are broken down to smaller ones by impellers and dispersed throughout the medium. This enables the creation of a uniform and homogeneous environment throughout the bioreactor.

Advantages of STRs

There are many advantages of STRs over other types. These include the efficient gas transfer to growing cells, good mixing of the contents and flexible operating conditions, besides the commercial availability of the bioreactors.

Bubble Column Bioreactors

In the bubble column bioreactor, the air or gas is introduced at the base of the column through perforated pipes or plates, or metal micro porous spargers. The flow rate of the air/gas influences the performance factors —O_2 transfer, mixing. The bubble column bioreactors may be fitted with perforated plates to improve performance. The vessel used for bubble column bioreactors is usually cylindrical with an aspect ratio of 4-6 (i.e., height to diameter ratio).

Airlift Bioreactors

In the airlift bioreactors, the medium of the vessel is divided into two interconnected zones by means of a baffle or draft tube. In one of the two zones referred to a riser, the air/gas is pumped. The other zone that receives no gas is the down comer. The dispersion flows up the riser zone while the down flow occurs in the down comer. There are two types of airlift bioreactors.

Internal-loop airlift bioreactor has a single container with a central draft tube that creates interior liquid circulation channels. These bioreactors are simple in design, with volume and circulation at a fixed rate for fermentation.

External loop airlift bioreactor possesses an external loop so that the liquid circulates through separate independent channels. These reactors can be suitably modified to suit the requirements of different fermentations. In general, the airlift bioreactors are more efficient than bubble columns, particularly for more denser suspensions of microorganisms. This is mainly because in these bioreactors, the mixing of the contents is better compared to bubble columns.

Airlift bioreactors are commonly employed for aerobic bioprocessing technology. They ensure a controlled liquid flow in a recycle system by pumping. Due to high efficiency, airlift bioreactors are sometimes preferred e.g., methanol production, waste water treatment, single-cell protein production. In general, the performance of the airlift bioreactors is dependent on the pumping (injection) of air and the liquid circulation.

Two-stage Airlift Bioreactors

Two-stage airlift bioreactors are used for the temperature dependent formation of products. Growing cells from one bioreactor (maintained at temperature 30 °C) are pumped into another bioreactor (at temperature 42 °C). There is a necessity for the two-stage airlift bioreactor, since it is very difficult to raise the temperature quickly from 30 °C to 42 °C in the same vessel. Each one of the bioreactors is fitted with valves and they are connected by a transfer tube and pump. The cells are grown in the first bioreactor and the bioprocess proper takes place in the second reactor.

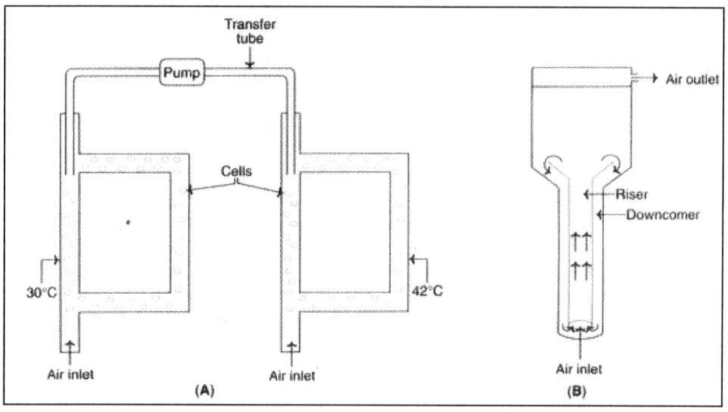

Type of bioreactors: (A) Two-Stage Airlift bioreactor (B) Tower bioreactor.

Tower Bioreactors

A pressure-cycle fermenter with large dimensions constitutes a tower bioreactor. A high hydrostatic pressure generated at the bottom of the reactor increases the solubility of O_2 in the medium. At the top of the riser, (with expanded top) reduces pressure and facilitates expulsion of CO_2. The medium flows back in the down comer and completes the cycle. The advantage with tower bioreactor is that it has high aeration capacities without having moving parts.

Fluidized Bed Bioreactors

Fluidized bed bioreactor is comparable to bubble column bioreactor except the top position is expanded to reduce the velocity of the fluid. The design of the fluidized bioreactors (expanded top and narrow reaction column) is such that the solids are retained in the reactor while the liquid flows out. These bioreactors are suitable for use to carry out reactions involving fluid suspended biocatalysts such as immobilized enzymes, immobilized cells, and microbial flocs.

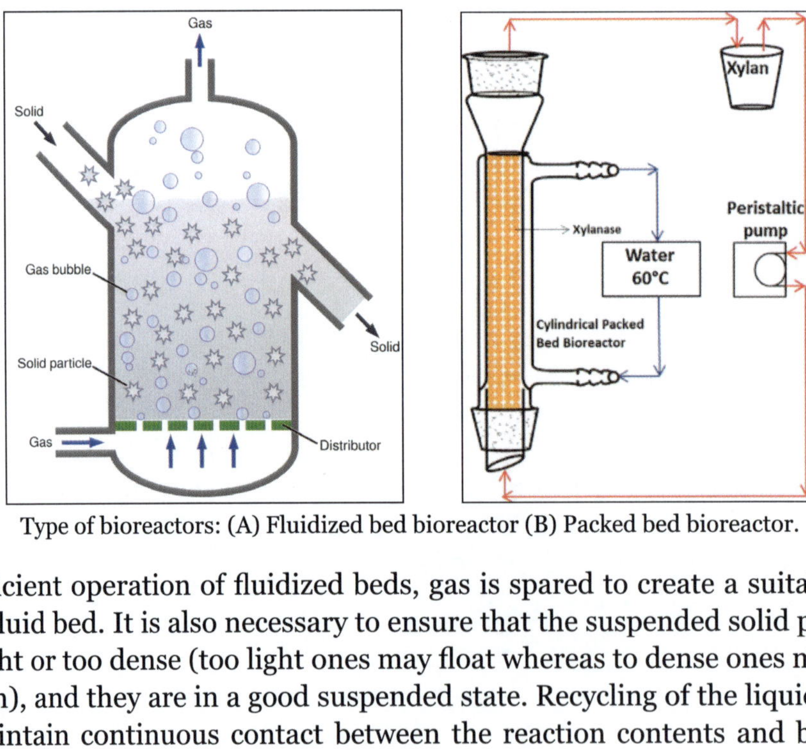

Type of bioreactors: (A) Fluidized bed bioreactor (B) Packed bed bioreactor.

For an efficient operation of fluidized beds, gas is spared to create a suitable gas-liquid-solid fluid bed. It is also necessary to ensure that the suspended solid particles are not too light or too dense (too light ones may float whereas to dense ones may settle at the bottom), and they are in a good suspended state. Recycling of the liquid is important to maintain continuous contact between the reaction contents and biocatalysts. This enable good efficiency of bioprocessing.

Packed Bed Bioreactors

A bed of solid particles, with biocatalysts on or within the matrix of solids, packed in a column constitutes a packed bed bioreactor. The solids used may be porous or non-porous gels, and they may be compressible or rigid in nature. A nutrient broth flows continuously over the immobilised biocatalyst. The products obtained in the packed bed bioreactor are released into the fluid and removed. While the flow of the fluid can be upward or downward, down flow under gravity is preferred.

The concentration of the nutrients (and therefore the products formed) can be increased by increasing the flow rate of the nutrient broth. Because of poor mixing, it is rather difficult to control the pH of packed bed bioreactors by the addition of acid or alkali. However, these bioreactors are preferred for bioprocessing technology involving product-inhibited reactions. The packed bed bioreactors do not allow accumulation of the products to any significant extent.

Photo-bioreactors

These are the bioreactors specialised for fermentation that can be carried out either by exposing to sunlight or artificial illumination. Since artificial illumination is expensive, only the outdoor photo-bioreactors are preferred. Certain important compounds are produced by employing photo-bioreactors e.g., p-carotene, asthaxanthin.

They are made up of glass or more commonly transparent plastic. The array of tubes or flat panels constitute light receiving systems (solar receivers). The culture can be circulated through the solar receivers by methods such as using centrifugal pumps or airlift pumps. It is essential that the cells are in continuous circulation without forming sediments. Further adequate penetration of sunlight should be maintained. The tubes should also be cooled to prevent rise in temperature.

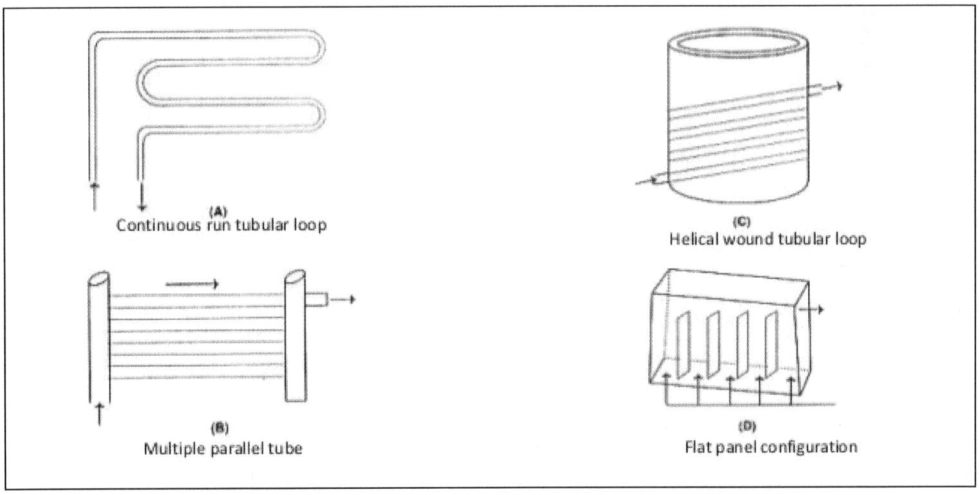

Types of photobioreactors (A) continuos run tabular loop
(B)Multiple parallel tube (C) Helical wound tabular loop (D) Flat panel configuration.

Photo-bioreactors are usually operated in a continuous mode at a temperature in the range of 25-40 °C. Microalgae and cyanobacteria are normally used. The organisms grow during day light while the products are produced during night.

BIOCONVERSION

Bioconversion, also known as *biotransformation*, is the conversion of organic materials, such as plant or animal waste, into usable products or energy sources by biological processes or agents, such as certain microorganisms. One example is the industrial production of cortisone, which one step is the bioconversion of progesterone to 11-alpha-Hydroxyprogesterone by *Rhizopus nigricans*. Another example is the bioconversion of glycerol to 1,3-propanediol, which is part of scientific research for many decades.

Another example of bioconversion is the conversion of organic materials, such as plant or animal waste, into usable products or energy sources by biological processes or agents, such as certain microorganisms, some detritivores or enzymes.

In the USA, the Bioconversion Science and Technology group performs multidisciplinary R&D for the Department of Energy's (DOE) relevant applications of bioprocessing, especially with biomass. Bioprocessing combines the disciplines of chemical engineering, microbiology and biochemistry. The Group 's primary role is investigation of the use of microorganism, microbial consortia and microbial enzymes in bioenergy research. New cellulosic ethanol conversion processes have enabled the variety and volume of feedstock that can be bioconverted to expand rapidly. Feedstock now includes materials derived from plant or animal waste such as paper, auto-fluff, tires, fabric, construction materials, municipal solid waste (MSW), sludge, sewage, etc.

Three Different Processes for Bioconversion

1. Enzymatic hydrolysis: A single source of feedstock, switchgrass for example, is mixed with strong enzymes which convert a portion of cellulosic material into sugars which can then be fermented into ethanol. Genencor and Novozymes are two companies that have received United States government Department of Energy funding for research into reducing the cost of cellulase, a key enzyme in the production cellulosic ethanol by this process.

2. Synthesis gas fermentation: A blend of feedstock, not exceeding 30% water, is gasified in a closed environment into a syngas containing mostly carbon monoxide and hydrogen. The cooled syngas is then converted into usable products through exposure to bacteria or other catalysts. BRI Energy, LLC is a company whose pilot plant in Fayetteville, Arkansas is currently using synthesis gas fermentation to convert a variety of waste into ethanol. After gasification, anaerobic bacteria (*Clostridium ljungdahlii*) are used to convert the syngas (CO, CO_2, and H_2) into ethanol. The heat generated by gasification is also used to co-generate excess electricity.

3. C.O.R.S. and Grub Composting: They are sustainable technologies that employ organisms that feed on organic matter to reduce and convert organic waste in to high quality feedstuff and oil rich material for the biodiesel industry. Organizations

pioneering this novel approach to waste management are EAWAG, ESR International, Prota Culture and BIOCONVERSION that created the *e*-CORS system to meet large scale organic waste management needs and environmental sustainability in both urban and livestock farming reality. This type of engineered system introduces a substantial innovation represented by the automatic modulation of the treatment, able to adapt conditions of the system to the biology of the scavenger used, improving their performances and the power of this technology.

ENERGY CROP

Energy crops are low-cost and low-maintenance crops grown solely for energy production (not for food). The crops are processed into solid, liquid or gaseous fuels, such as pellets, bioethanol or biogas. The fuels are burned to generate power or heat.

The plants are generally categorized as woody or herbaceous. Woody plants include willow and poplar, herbaceous plants include *Miscanthus x giganteus* and *Pennisetum purpureum* (both known as elephant grass). Herbaceous crops, while physically smaller than trees, store roughly twice the amount of CO_2 (in the form of carbon) below ground, compared to woody crops.

Through biotechnological procedures such as genetic modification plants can be manipulated to create higher yields. Relatively high yields can also be realized with existing cultivars. However, some additional advantages such as reduced associated costs (i.e. costs during the manufacturing process) and less water use can only be accomplished by using genetically modified crops.

CO_2 Neutrality

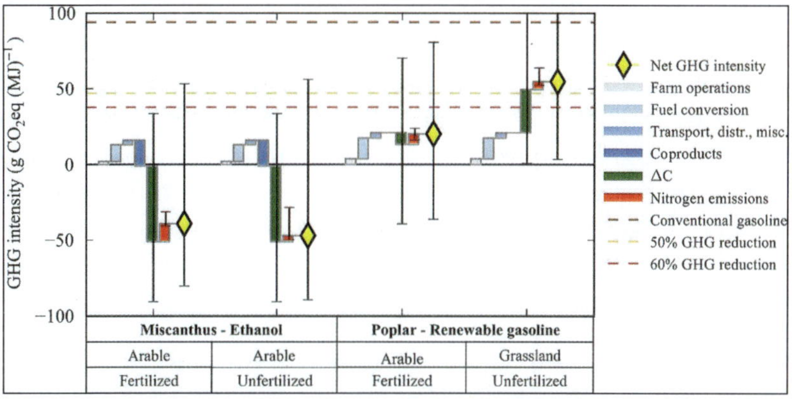

GHG/CO_2/carbon negativity for Miscanthus x giganteus production pathways.

The amount of carbon sequestrated and the amount of GHG (greenhouse gases) emitted will determine if the total GHG life cycle cost of a bio-energy project is positive, neutral or negative. Specifically, a GHG/carbon negative life cycle is possible if the total below-ground carbon accumulation more than compensates for the above-ground total life-cycle GHG emissions. Whitaker et al. estimates that for Miscanthus x giganteus, carbon neutrality and even negativity is within reach. Basically, the yield and related carbon sequestration is so high that it more than compensates for both farm operations emissions, fuel conversion emissions and transport emissions. The graphic on the right displays two CO_2 negative Miscanthus x giganteus production pathways, represented in gram CO_2-equivalents per megajoule. The yellow diamonds represent mean values.

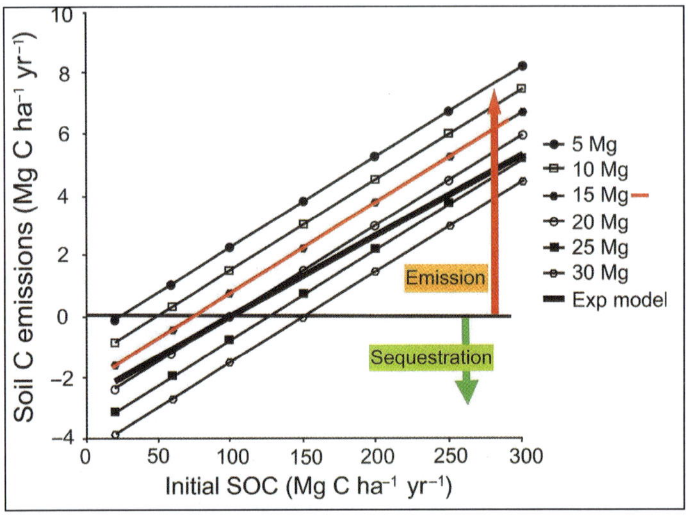

Relationship between above-ground yield (diagonal lines), soil organic carbon (X axis), and soil's potential for successful/unsuccessful carbon sequestration (Y axis). Basically, the higher the yield, the more land is usable as a GHG mitigation tool (including relatively carbon rich land.)

One should note that successful sequestration is dependent on planting sites, as the best soils for sequestration are those that are currently low in carbon. The varied results displayed in the graph highlights this fact. Milner et al. argues that for the UK, successful sequestration is expected for arable land over most of England and Wales, with unsuccessful sequestration expected in parts of Scotland, due to already carbon rich soils (existing woodland). Also, for Scotland, the relatively lower yields in this colder climate makes CO_2 negativity harder to achieve. Soils already rich in carbon includes peatland and mature forest. Grassland can also be carbon rich, however Milner et al. further argues that the most successful carbon sequestration in the UK takes place below improved grasslands. The bottom graphic displays the estimated yield necessary to achieve CO_2 negativity for different levels of existing soil carbon saturation.

The perennial rather than annual nature of Miscanthus crops implies that the significant below-ground carbon accumulation each year is allowed to continue undisturbed. No annual plowing or digging means no increased carbon oxidation and no stimulation of the microbe populations in the soil, and therefore no accelerated carbon-to-CO_2 conversion happening in the soil every spring.

Types

Solid Biomass

Elephant grass (Miscanthus giganteus) is an experimental energy crop.

Solid biomass, often pelletized, is used for combustion in thermal power stations, either alone or co-fired with other fuels. Alternatively it may be used for heat or combined heat and power (CHP) production.

In short rotation coppice (SRC) agriculture, fast growing tree species like willow and poplar are grown and harvested in short cycles of three to five years. These trees grow best in wet soil conditions. An influence on local water conditions can not be excluded. Establishment close to vulnerable wetland should be avoided.

Gas Biomass (Methane)

Whole crops such as maize, Sudan grass, millet, white sweet clover, and many others can be made into silage and then converted into biogas. Anaerobic digesters or biogas plants can be directly supplemented with energy crops once they have been ensiled into silage. The fastest growing sector of German biofarming has been in the area of "Renewable Energy Crops" on nearly 500,000 ha (1,200,000 acres) of land (2006). Energy crops can also be grown to boost gas yields where feedstocks have a low energy content, such as manures and spoiled grain. It is estimated that the energy yield presently of bioenergy crops converted via silage to methane is about 2 GWh/km^2 (1.8×10^{10} BTU/sq mi) annually. Small mixed cropping enterprises with animals can use a portion of their acreage to grow and convert energy crops and sustain the entire farms energy requirements with about one fifth of the acreage. In Europe and especially Germany, however, this rapid growth has occurred only with substantial government support, as in the German bonus system for renewable energy. Similar developments of integrating crop farming and bioenergy production via silage-methane have been almost entirely overlooked in N. America, where political and structural issues and a huge continued push to centralize energy production has overshadowed positive developments.

Liquid Biomass

Biodiesel

Coconuts sun-dried in Kozhikode, Kerala for making copra, the dried meat, or kernel, of the coconut. Coconut oil extracted from it has made copra an important agricultural commodity for many coconut-producing countries. It also yields coconut cake which is mainly used as feed for livestock.

Pure biodiesel (B-100), made from soybeans.

European production of biodiesel from energy crops has grown steadily in the last decade, principally focused on rapeseed used for oil and energy. Production of oil/biodiesel from rape covers more than 12,000 km^2 in Germany alone, and has doubled in the past 15 years. Typical yield of oil as pure biodiesel may be is 100,000 L/km^2 (68,000 US gal/sq mi; 57,000 imp gal/sq mi) or more, making biodiesel crops economically attractive, provided sustainable crop rotations exist that are nutrient-balanced and preventative of the spread of disease such as clubroot. Biodiesel yield of soybeans is significantly lower than that of rape.

Typical oil extractable by weight	
Crop	Oil %
Copra	62
Castor seed	50
Sesame	50
Groundnut kernel	42

Jatropha	40
Rapeseed	37
Palm kernel	36
Mustard seed	35
Sunflower	32
Palm fruit	20
Soybean	14
Cotton seed	13

Bioethanol

Energy crops for biobutanol are grasses. Two leading non-food crops for the production of cellulosic bioethanol are switchgrass and giant miscanthus. There has been a preoccupation with cellulosic bioethanol in America as the agricultural structure supporting biomethane is absent in many regions, with no credits or bonus system in place. Consequently, a lot of private money and investor hopes are being pinned on marketable and patentable innovations in enzyme hydrolysis and the like.

Bioethanol also refers to the technology of using principally corn (maize seed) to make ethanol directly through fermentation, a process that under certain field and process conditions can consume as much energy as is the energy value of the ethanol it produces, therefore being non-sustainable. New developments in converting grain stillage (referred to as distillers grain stillage or DGS) into biogas energy looks promising as a means to improve the poor energy ratio of this type of bioethanol process.

Human beings run on energy and to run our bodies, we use food as fuel. It may seem ironic, then, that in a bid to harness more kinds of natural and renewable energy, certain plants are grown with the single-minded agenda of turning them into biofuel or combusting them to generate energy. Such plants are dubbed 'energy crops.'

Short Rotation Energy Crops

There are two kinds of short rotation energy crops, depending on the cycle and scale. In short rotation coppice (SRC), fast-growing young trees are cut at the stump every winter when they lie dormant, thus giving rise to many new stems in the growing season (the yield stands at 2-4 years). Trees like poplar and willow are popular choices for SRC. Short Rotation forestry is planting a site and then felling the tree when their stems reach a diameter of 10-20 cm, at chest-height. This cycle is longer and takes place every 8-20 years; popular plants include sycamore, ash, poplar, eucalyptus and beech.

Non-Woody Crops and Grasses

Miscanthus is the most popular of all non-woody energy crops and gives a good annual yield as opposed to short rotation energy crops that give yield only every 2-20 years. Other potential energy crops include hemp and varieties of reed, rye and switchgrass (from the grasses family) too can be considered. With these plants though, there's always the threat of invasion so these need to be taken up with care and caution.

Agricultural Energy Crops

By virtue of being high in carbon content, many conventional crops such as sugar crops (sugar beet), starch crops (wheat, maize, and potatoes) and oil crops (rapeseed, waste vegetable oil) are either used straight as fuel or hydrolyzed into biofuel.

Aquatic Crops (Hydroponics)

The advantage of aquatic crops is that they don't use land and they take all that they need in form of nutrients from the water, thus being excellent in photosynthesis. Algae, both microscopic and macroscopic (such as seaweed, kelp) and other pond and lake flora are good forms of energy crops – the disadvantage lies in their high water content which needs to be dried up before being used as biomass.

Advantages of Energy Crops

Growing agriculture for energy means lesser dependence on fossil fuels and thus a decrease in mining and exports. Another big advantage is that biomass is freely available. Growing plants or a plantation means that they use up carbon dioxide and release oxygen, thus leading to a more oxygenated atmosphere. In developed countries where there are huge landfills built to handle agricultural waste, biomass energy could come in handy, thus freeing that land for other and more beneficial purposes. The government also provides many grants for farmers interested in the same.

Disadvantages of Energy Crops

The land used for the agriculture of biomass may in fact be needed to grow consumable crops – or even as land for housing or for recreational or commercial purposes. Also, to be able to sustain energy crops on a large scale, the costs associated with the conversion of the biomass into fuel need to be brought down.

Lastly the one big reason that makes energy crops stand at a disadvantage is that sometimes the conversion causes environmental pollution.

ENERGY FORESTRY

Energy forestry is a form of forestry in which a fast-growing species of tree or woody shrub is grown specifically to provide biomass or biofuel for heating or power generation.

The two forms of energy forestry are short rotation coppice and short rotation forestry:

- Short rotation coppice may include tree crops of poplar, willow or eucalyptus, grown for two to five years before harvest.
- Short rotation forestry are crops of alder, ash, birch, eucalyptus, poplar, and sycamore, grown for eight to 20 years before harvest.

Benefits

The main advantage of using "grown fuels", as opposed to fossil fuels such as coal, natural gas and oil, is that while they are growing they absorb the near-equivalent in carbon dioxide (an important greenhouse gas) to that which is later released in their burning. In comparison, burning fossil fuels increases atmospheric carbon unsustainably, by using carbon that was added to the Earth's carbon sink millions of years ago. This is a prime contributor to climate change.

According to the FAO, compared to other energy crops, wood is among the most efficient sources of bioenergy in terms of quantity of energy released by unit of carbon emitted. Other advantages of generating energy from trees, as opposed to agricultural crops, are that trees do not have to be harvested each year, the harvest can be delayed when market prices are down, and the products can fulfil a variety of end-uses.

Yields of some varieties can be as high as 12 oven dry tonnes every year. However, commercial experience on plantations in Scandinavia have shown lower yield rates.

These crops can also be used in bank stabilisation and phytoremediation. In fact, experiments in Sweden with willow plantations have proved to have many beneficial effects on the soil and water quality when compared to conventional agricultural crops (such as cereal).

Problems

Although in many areas of the world government funding is still required to support large scale development of energy forestry as an industry, it is seen as a valuable component of the renewable energy network and will be increasingly important in the future.

Growing Trees is Relatively Water Intensive. The system of energy forestry has faced criticism over food vs. fuel, whereby it has become financially profitable to replace food crops with energy crops. It has to be noted, however, that such energy forests do not necessarily compete with food crops for highly productive land as they can be grown on slopes, marginal, or degraded land as well - sometimes even with long-term restoration purposes in mind.

BIO-ENERGY WITH CARBON CAPTURE AND STORAGE

Bio-energy with carbon capture and storage (BECCS) is the process of extracting bioenergy from biomass and capturing and storing the carbon, thus removing it from the atmosphere.

The carbon in the trees or crops used for the biomass comes from the greenhouse gas carbon dioxide (CO_2), which they extract from the atmosphere whilst growing. As of 2019 most BECCS stores some of the carbon and leaves the rest with the bioenergy in the form of a biofuel, such as ethanol. By providing aviation biofuel and reducing the environmental impact of aviation BECCS may help with climate change mitigation. Alternatively the bioenergy could be used in the form of electricity and a larger proportion of the CO_2 stored permanently by injection into geological formations.

As of 2019 five facilities around the world are actively using BECCS technologies and are capturing approximately 1.5 million tonnes per year (Mtpa) of CO_2 The IPCC Fifth Assessment Report by the Intergovernmental Panel on Climate Change (IPCC), implies a range of negative emissions from BECCS of 0 to 22 gigatonnes.

There are other non-BECCS forms of carbon dioxide removal and storage including afforestation, biochar, carbon dioxide air capture and biomass burial and enhanced weathering. Pyrogenic carbon capture and storage (PyCCS) or biochar is superior in fixing carbon for a longer time. Carbon dioxide injected into geologic formations eventually leaks back into the atmosphere due to seismic activity and natural faults and problems with the seal of the ancient injection pits.

Negative Emission

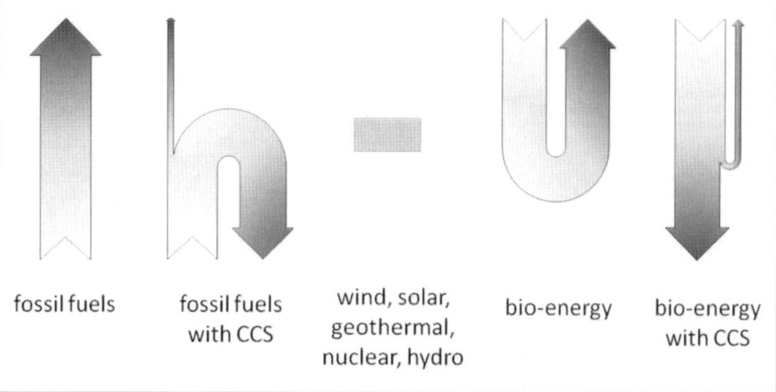

Carbon flow schematic for different energy systems.

The main appeal of BECCS is in its ability to result in negative emissions of CO_2. The capture of carbon dioxide from bioenergy sources effectively removes CO_2 from the atmosphere.

Bio-energy is derived from biomass which is a renewable energy source and serves as a carbon sink during its growth. During industrial processes, the biomass combusted or processed re-releases the CO_2 into the atmosphere. The process thus results in a net zero emission of CO_2, though this may be positively or negatively altered depending on the carbon emissions associated with biomass growth, transport and processing, Carbon capture and storage (CCS) technology serves to intercept the release of CO_2 into the atmosphere and redirect it into geological storage locations. CO_2 with a biomass origin is not only released from biomass fuelled power plants, but also during the production of pulp used to make paper and in the production of biofuels such as biogas and bioethanol. The BECCS technology can also be employed on such industrial processes.

It is argued that through the BECCS technology, carbon dioxide is trapped in geologic formations for very long periods of time, whereas for example a tree only stores its carbon during its lifetime. In its report on the CCS technology, IPCC projects that more than 99% of carbon dioxide which is stored through geologic sequestration is likely to stay in place for more than 1000 years. While other types of carbon sinks such as the ocean, trees and soil may involve the risk of adverse feedback loops at increased temperatures, BECCS technology is likely to provide a better permanence by storing CO_2 in geological formations.

The amount of CO_2 that has been released to date is believed to be too much to be able to be absorbed by conventional sinks such as trees and soil in order to reach low emission targets. In addition to the presently accumulated emissions, there will be significant additional emissions during this century, even in the most ambitious low-emission scenarios. BECCS has therefore been suggested as a technology to reverse the emission trend and create a global system of net negative emissions. This implies that the emissions would not only be zero, but negative, so that not only the emissions, but the absolute amount of CO_2 in the atmosphere would be reduced.

Application

Source	CO_2 Source	Sector
Ethanol production	Fermentation of biomass such as sugarcane, wheat or corn releases CO_2 as a by-product.	Industry
Pulp and paper mills	• CO_2 produced in recovery boilers, • CO_2 produced in lime kilns, • For gasification technologies, CO_2 is produced during the gasification of black liquor and biomass such as the tree bark and wood, • Huge amounts of CO_2 are also released by the combustion of syngas, a product of gasification, in the combined cycle process.	Industry
Biogas production	In the biogas upgrading process, CO_2 is separated from the methane to produce a higher quality gas.	Industry
Electrical power plants	Combustion of biomass or biofuel in steam or gas powered generators releases CO_2 as a by-product.	Energy
Heat power plants	Combustion of biofuel for heat generation releases CO_2 as a by-product. Usually used for district heating.	Energy

Cost

The IPCC states that estimations for BECCS cost range from \$60-\$250 per ton of CO_2.

Technology

The main technology for CO_2 capture from biotic sources generally employs the same technology as carbon dioxide capture from conventional fossil fuel sources. Broadly, three different types of technologies exist: post-combustion, pre-combustion, and oxy-fuel combustion.

Oxy-combustion

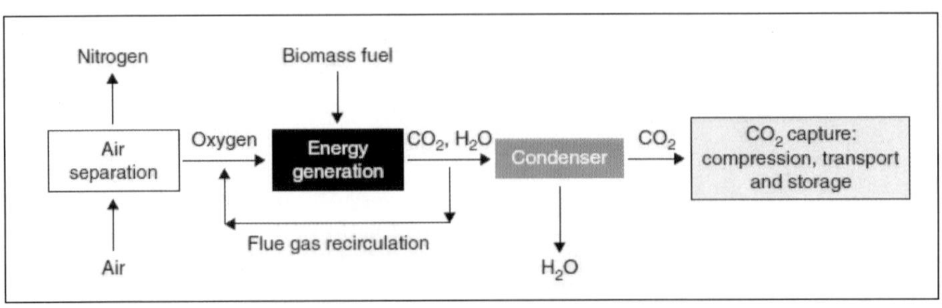

Overview of oxy-fuel combustion for carbon capture from biomass, showing the key processes and stages; some purification is also likely to be required at the dehydration stage.

Oxy-fuel combustion has been a common process in the glass, cement and steel industries. It is also a promising technological approach for CCS. In oxy-fuel combustion, the main difference from conventional air firing is that the fuel is burned in a mixture of O_2 and recycled flue gas. The O_2 is produced by an air separation unit (ASU), which removes the atmospheric N_2 from the oxidizer stream. By removing the N_2 upstream of the process, a flue gas with a high concentration of CO_2 and water vapor is produced, which eliminates the need for a post-combustion capture plant. The water vapor can be removed by condensation, leaving a product stream of relatively high-purity CO_2 which, after subsequent purification and dehydration, can be pumped to a geological storage site.

The key challenges of BECCS implementation using oxy-combustion method is associated with combustion process. For the high volatile content biomass, the mill temperature has to be maintain at low temperature to prevent the risk of fire and explosion. In addition, the flame temperature is lower. Therefore, the concentration of oxygen needs to be increased up to 27-30%.

Pre-combustion

Pre-combustion carbon capture refers to the process of capturing CO_2 before the energy generations. It is often composed of five operating stations: oxygen generation, syngas generation, CO_2 separation, CO_2 compression, and power generation. Basically, the fuel will first go through a gasification process by reacting with oxygen to form a stream of CO and H_2, which is syngas. The products will then go through a water-gas shift reactor to form CO_2 and H_2. The CO_2 that is produced will then be captured, and the H_2, which is a clean source, will be used for combustion to generate energy. The process of gasification combined with syngas production is called Integrated Gasification Combined Cycle (IGCC). Normally, it would require an Air Separation Unit (ASU) to serve as the oxygen source. However, research proves that with the same flue gas, oxygen gasification has little advantage over air gasification, and they both have a similar thermal efficiency of roughly 70% using coal as the fuel source. Thus, the use of ASU is not really necessary in pre-combustion.

Biomass is considered "sulfur-free" as a fuel for the pre-combustion capture. However, there are other trace elements in biomass combustion such as K and Na that could accumulate in the system and finally cause the degradation of the mechanical parts. Thus, further developments of the separation techniques for those trace elements are needed. And also, after the gasification process, CO_2 takes up to 13% - 15.3% by mass in the syngas stream for biomass sources, while it is only 1.7% - 4.4% for coal. This limit the conversion of CO to CO_2 in the water gas shift, and the production rate for H_2 will decrease accordingly. However, the thermal efficiency of the pre-combustion capture using biomass resembles that of coal which is around 62% - 100%. Besides, some researches also proved that instead of using a biomass/water slurry fuel feed, using a dry system is more thermally efficient and also more practical for biomass.

Post-combustion

In addition to pre-combustion and oxy-fuel combustion technologies, post-combustion is a promising technology which can be used to extract CO_2 emission from biomass fuel resources. During the process, CO_2 is separated from the other gases in the flue gas stream after the biomass fuel is burnt and undergo separation process. Because it has the ability to be retrofitted to some existing power plants such as steam boilers or other newly built power stations, post-combustion technology is considered as a better option than pre-combustion technology. According to the fact sheets *U.S. Consumption of bio-energy with carbon capture and storage* released in March 2018, the efficiency of post-combustion technology is expected to be 95% while pre-combustion and oxy-combustion capture CO_2 at an efficient rate of 85% and 87.5% respectively.

Development for current post-combustion technologies has not been entirely done due to several problems. One of the major concerns using this technology to capture carbon dioxide is the parasitic energy consumption. If the capacity of the unit is designed to be small, the heat loss to the surrounding is great enough to cause to many negative consequences. Another challenge of post-combustion carbon capture is how to deal with the mixture's components in the flue gases from initial biomass materials after combustion. The mixture consists of a high amount of alkali metals, halogens, acidic elements, and transition metals which might have negative impacts on the efficiency of the process. Thus, the choice of specific solvents and how to manage the solvent process should be carefully designed and operated.

Biomass Feedstocks

Biomass sources used in BECCS include agricultural residues & waste, forestry residue & waste, industrial & municipal wastes, and energy crops specifically grown for use as fuel. Current BECCS projects capture CO_2 from ethanol bio-refinery plants and municipal solid waste (MSW) recycling center.

At Ethanol Plants

Illinois Carbon Capture and Storage (IL-CCS) is one of the milestones, being the first industrial-scaled BECCS project, in the early 21st century. Located in Decatur, Illinois, USA, IL-CCS captures CO_2 from Archer Daniels Midland (ADM) ethanol plant. The captured CO_2 is then injected under the deep saline formation at Mount Simon Sandstone. IL-CCS consists of 2 phases. The first being a pilot project which was implemented from 11/2011 to 11/2014. Phase 1 has a capital cost of around 84 million US dollars. Over the 3-year period, the technology successfully captured and sequestered 1million tonne of CO_2 from the ADM plant to the aquifer. No leaking of CO_2 from the injection zone was found during this period. The project is still being monitored for future reference. The success of phase 1 motivated the deployment of phase 2, bringing IL-CCS (and BECCS) to industrial scale. Phase 2 has been in operation since 11/2017 and also use the same injection zone at Mount Simon Sandstone as phase 1. The capital cost for second phase is about 208 million US dollars including 141 million US dollar fund from the Department of Energy. Phase 2 has capturing capacity about 3 time larger than the pilot project (phase 1). Annually, IL-CCS can capture mourned 1 million tonne of CO_2. With the largest of capturing capacity, IL-CCS is currently the largest BECCS project in the world.

In addition to the IL-CCS project, there are about three more projects that capture CO_2 from the ethanol plant at smaller scales. For example, Arkalon at Kansas, USA can capture 0.18-0.29 $MtCO_2$/yr, OCAP at Netherlands can capture about 0.1-0.3 $MtCO_2$/yr, and Husky Energy at Canada can capture 0.09-0.1 $MtCO_2$/yr.

At MSW Recycling Centers

Beside capturing CO_2 from the ethanol plants, currently, there are 2 models in Europe are designed to capture CO_2 from the processing of Municipal Solid Waste. The Klemetsrud Plant at Oslo, Norway use biogenic municipal solid waste to generate 175 GWh and capture 315 Ktonne of CO_2 each year. It uses absorption technology with Aker Solution Advanced Amine solvent as a CO_2 capture unit. Similarly, the ARV Duiven at Netherlands uses the same technology, but it captures less CO_2 than the previous model. ARV Duiven generates around 126 GWh and only capture 50 Ktonne of CO_2 each year.

Techno-economics of BECCS and the TESBiC Project

The largest and most detailed techno-economic assessment of BECCS was carried out by cmcl innovations and the TESBiC group (Techno-Economic Study of Biomass to CCS) in 2012. This project recommended the most promising set of biomass fueled power generation technologies coupled with carbon capture and storage (CCS). The project outcomes lead to a detailed "biomass CCS roadmap" for the U.K..

Challenges

Environmental Considerations

Some of the environmental considerations and other concerns about the widespread implementation of BECCS are similar to those of CCS. However, much of the critique towards CCS is that it may strengthen the dependency on depletable fossil fuels and environmentally invasive coal mining. This is not the case with BECCS, as it relies on renewable biomass. There are however other considerations which involve BECCS and these concerns are related to the possible increased use of biofuels. Biomass production is subject to a range of sustainability constraints, such as: scarcity of arable land and fresh water, loss of biodiversity, competition with food production, deforestation and scarcity of phosphorus. It is important to make sure that biomass is used in a way that maximizes both energy and climate benefits. There has been criticism to some suggested BECCS deployment scenarios, where there would be a very heavy reliance on increased biomass input.

Large areas of land would be required to operate BECCS on an industrial scale. To remove 10 billion tons of CO_2, upwards of 300 million hectares of land area (larger than India) would be required. As a result, BECCS risks using land that could be better suited to agriculture and food production, especially in developing countries.

These systems may have other negative side effects. There is however presently no need to expand the use of biofuels in energy or industry applications to allow for BECCS deployment. There is already today considerable emissions from point sources of biomass derived CO_2, which could be utilized for BECCS. Though, in possible future bio-energy system upscaling scenarios, this may be an important consideration.

Upscaling BECCS would require a sustainable supply of biomass - one that does not challenge our land, water, and food security. Using bio-energy crops as feedstock will not only cause sustainability concerns but also require the use of more fertilizer leading to soil contamination and water pollution. Moreover, crop yield is generally subjected to climate condition, i.e. the supply of this bio-feedstock can be hard to control. Bioenergy sector must also expand to meet the supply level of biomass. Expanding bioenergy would require technical and economic development accordingly.

Technical Challenges

Just as other carbon capture and storage technologies, one of the challenges of applying BECCS technology is to find suitable geographic locations to build combustion plant and to sequester captured CO_2. If biomass sources are not close by the combustion unit, transporting biomass emits CO_2 offsetting the amount of CO_2 captured by BECCS. BECCS also face technical concerns about efficiency of burning biomass. While each type of biomass has a different heating value, biomass in general is a low-quality fuel. Thermal conversion of biomass has a typical efficiency of 20-27%. Coal-fired plant has an efficiency of about 37% for comparison.

BECCS also faces a question whether the process is actually energy positive. Low energy conversion efficiency, energy-intensive biomass supply, combined with the energy required to power the CO_2 capture and storage unit impose energy penalty on the system. This might lead to a low power generation efficiency.

Alternative Biomass Sources

Agricultural and Forestry Residues

Globally, 14 Gt of forestry residue and 4.4 Gt residues from crop production (mainly barley, wheat, corn, sugarcane and rice) are generated every year. This is a significant amount of biomass which can be combusted to generate 26 EJ/year and achieve a 2.8 Gt of negative CO_2 emission through BECCS. Utilizing residues for carbon capture will provide social and economic benefits to rural communities. Using waste from crops and forestry is a way to avoid the ecological and social challenges of BECCS.

Municipal Solid Waste

Municipal solid waste(MSW) is one of the newly developed sources of biomass. Two current BECCS plants are using MSW as feedstocks. Waste collected from daily life is recycled via incineration waste treatment process. Waste goes through high temperature thermal treatment and the heat generated from combusting organic part of waste is used to generate electricity. CO_2 emitted from this process is captured through absorption using MEA. For every 1 kg of waste combusted, 0.7 kg of negative CO_2 emission is achieved. Utilizing solid waste also have other environmental benefits.

Co-firing Coal with Biomass

There are currently 200 cofiring plants in the world, including 40 in the US. Studies showed that by mixing coal with biomass, we could reduce the amount of CO_2 emitted. The concentration of CO_2 in the flue gas is an important key to determine the efficiency of CO_2 capture technology. The concentration of CO_2 in the flue gas from the co-firing power plant is roughly the same as coal plant, about 15% . This means that we can reduce our reliance on fossil fuel.

Even though co-firing will have some energy penalty, it still offers higher net efficiency than the biomass combustion plants. Co-firing biomass with coal will result more energy production with less input material. Currently, the modern 500 MW coal power plant can take up to 15% biomass without changing the component of the steam boiler. This promising potential allows co-firing power plant become more favorable than dedicated bio-electricity.

It is estimated that by replacing 25% of coal with biomass at existing power plant in China and the U.S, we can reduce emission by 1Gt per year. The amount of negative CO_2 emitted depends on the composition of coal and biomass. 10% biomass can reduce 0.5 Gt CO_2 per year and with 16% biomass can achieve zero emission. Direct-co-firing (20% biomass) give us negative emission of -26 kg CO_2/MWh (from 93 kg CO_2/MWh).

Biomass cofiring with coal has efficiency near those of coal combustion. Cofiring can be easily applied to existing coal-fired power plant at low cost. The implementation of co-firing power plant on the global scale is still a challenge. The biomass resources have to meet strictly the sustainability criteria and the co-firing project would need the support in term of economic and policy from the governments.

Even though co-firing plant may be an immediate contribution to solving the global warming and climate change issues, co-firing still has some challenges that need to consider. Due to the moisture content of biomass, it will affect the calorific value of the combustor. In addition, high volatile biomass would highly influence the reaction rate and the temperature of the reactor; especially, it may lead to the explosion of furnace.

Instead of co-firing, full conversion from coal to biomass of one or more generating units in a plant may be preferred.

ETHANOL FUEL ENERGY BALANCE

In order to create ethanol, all biomass needs to go through some of these steps: it needs to be grown, collected, dried, fermented, and burned. All of these steps require resources and an infrastructure. The ratio of the energy released by burning the resulting ethanol fuel to the energy used in the process, is known as the ethanol fuel energy balance (sometimes called "Net energy gain") and studied as part of the wider field of energy economics. It compiled in a 2007 National Geographic Magazine article point to modest results for corn ethanol produced in the US: 1 unit of energy input equals 1.3 energy units of corn ethanol energy. The energy balance for sugarcane ethanol produced in Brazil is much more favorable, 1 to 8. Over the years, however, many reports have been produced with contradicting energy balance estimates. A 2006 University of California Berkeley study, after analyzing six separate studies, concluded that producing ethanol from corn uses marginally less petroleum than producing gasoline.

Energy Balance Reports

In 1995 the USDA released a report stating that the net energy balance of corn ethanol in the United States was an average of 1.24. It was previously considered to have a negative net energy balance. However, due to increases in corn crop yield and more efficient farming practices corn ethanol had gained energy efficiency.

Ken Cassman, a professor of agronomy at the University of Nebraska–Lincoln, said in 2008 that ethanol has a substantial net positive direct energy balance—1.5 to 1.6 more units of energy are derived from ethanol than are used to produce it. Comparing 2008 to 2003, Alan Tiemann of Seward, a Nebraska Corn Board member, said that ethanol plants produce 15 percent more ethanol from a bushel of corn and use about 20 percent less energy in the process. At the same time, corn growers are more efficient, producing more corn per acre and using less energy to do so.

Opponents of corn ethanol production in the U.S. often quote the 2005 paper of David Pimentel, a retired Entomologist, and Tadeusz Patzek, a Geological Engineer from UC Berkeley. Both have been exceptionally critical of ethanol and other biofuels. Their studies contend that ethanol, and biofuels in general, are "energy negative", meaning they take more energy to produce than is contained in the final product.

A 2006 article in Science offers the consensus opinion that current corn ethanol technologies had similar greenhouse gas emissions to gasoline, but was much less petroleum-intensive than gasoline. Fossil fuels also require significant energy inputs which have seldom been accounted for in the past.

Ethanol is not the only product created during production. By-products also have energy content. Corn is typically 66% starch and the remaining 33% is not fermented. This unfermented component is called distillers grain, which is high in fats and proteins, and makes a good animal feed supplement.

In 2000, Dr. Michael Wang, of Argonne National Laboratory, wrote that these ethanol by-products are the most contentious issue in evaluating the energy balance of ethanol. He wrote that Pimentel assumes that corn ethanol entirely replaces gasoline and so the quantity of by-products is too large for the market to absorb, and they become waste. At lower quantities of production, Wang finds it appropriate to credit corn ethanol based on the input energy requirement of the feed product or good that the ethanol by-product displaces. In 2004, a USDA report found that co-products accounting made the difference between energy ratios of 1.06 and 1.67. In 2006, MIT researcher Tiffany Groode came to similar conclusions about the co-product issue.

In Brazil where sugar cane is used, the yield is higher, and conversion to ethanol is more energy efficient than corn. Recent developments with cellulosic ethanol production may improve yields even further.

In 2006 a study from the University of Minnesota found that corn-grain ethanol produced 1.25 units of energy per unit put in.

A 2008 study by the University of Nebraska found a 5.4 energy balance for ethanol derived specifically from switchgrass. This estimate is better than in previous studies and according to the authors partly due to the larger size of the field trial (3-9 ha) on 10 farms.

Variables

According to DoE, to evaluate the net energy of ethanol four variables must be considered:

1. The amount of energy contained in the final ethanol product.
2. The amount of energy directly consumed to make the ethanol (such as the diesel used in tractors).
3. The quality of the resulting ethanol compared to the quality of refined gasoline.
4. The energy indirectly consumed (in order to make the ethanol processing plant, etc.).

Much of the current academic discussion regarding ethanol currently revolves around issues of system borders. This refers to how complete of a picture is drawn for energy inputs. There is debate on whether to include items like the energy required to feed the people tending and processing the corn, to erect and repair farm fences, even the amount of energy a tractor represents.

In addition, there is no consensus on what sort of value to give the rest of the corn (such as the stalk), commonly known as the 'coproduct.' Some studies leave it on the field to protect the soil from erosion and to add organic matter, while others take and burn the coproduct to power the ethanol plant, but do not address the resulting soil erosion (which would require energy in the form of fertilizer to replace). Depending on the ethanol study you read, net energy returns vary from .7-1.5 units of ethanol per unit of fossil fuel energy consumed. For comparison, that same one unit of fossil fuel invested in oil and gas extraction (in the lower 48 States) will yield 15 units of gasoline, a yield an order of magnitude better than current ethanol production technologies, ignoring the energy quality arguments above and the fact that the gain (14 units) is both declining and not carbon neutral.

In this regard, geography is the decisive factor. In tropical regions with abundant water and land resources, such as Brazil and Colombia, the viability of production of ethanol from sugarcane is no longer in question; in fact, the burning of sugarcane residues (bagasse) generates far more energy than needed to operate the ethanol plants, and many of them are now selling electric energy to the utilities. However, while there may be a positive net energy return at the moment, recent research suggests that the sugarcane plantations are not sustainable in the long run, as they are depleting the soil of nutrients and carbon matter On the other hand, productivity of sugar cane per land area in Brazil has consistently grown over the decades; sugar cane has been shown to be less depleting to the soil than cattle and yearly cultures; and there are many regions in the country where sugar cane has been cultivated for centuries. Those facts suggest that related soil depletion processes are very slow and therefore ethanol from sugar cane may be far more sustainable in the long run than common fossil fuel alternatives. Besides, since the energy surplus is high in the case of sugar cane ethanol, conceivably part of that energy can be used to synthesize fertilizers and replenish soil depletion a long time, therefore making the process indefinitely sustainable.

The picture is different for other regions, such as most of the United States, where the climate is too cool for sugarcane. In the U.S., agricultural ethanol is generally obtained from grain, chiefly corn. But it can also be obtained from cellulose, a more energy balanced bioethanol.

Clean Production Bioethanol

Clean production bioethanol is a biofuel obtained by maximizing non-greenhouse gas emitting (renewable) resources:

- Energy directly consumed to make the ethanol is renewable energy. The farm equipment and ethanol plant use an ethanol engine, biodiesel, air engine or electricity cogenerated during ethanol production, or even wind power and solar energy.

- Energy indirectly consumed is, as much as possible, renewable. Examples would be reducing either the amount or fossil carbon content of applied pest control chemicals and fertilizers, or accomplishing deliveries of farm inputs or of finished bioethanol fuel to market that minimize the use of fossil fuels. Optimally located biomass and ethanol production must balance many factors: minimizing distances to and from markets, effectively collecting and employing biomass wastes, maximizing crop yields based on enduring soil quality, available natural pest control and adequate sun and water, and optimizing a sufficient mix and rotation of plant species on cultivated, fallow and preserved land for human, animal and energy consumption.

Using ethanol returns carbon to the atmosphere whereas burning gasoline adds carbon to the atmosphere. Thus the effects of gasoline burning increase over time.

PELLET STOVE

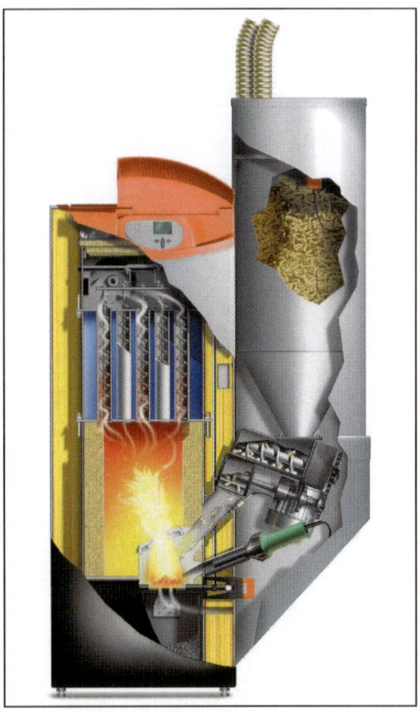

Modern pellet stove cross section.

A pellet stove is a stove that burns compressed wood or biomass pellets to create a source of heat for residential and sometimes industrial spaces. By steadily feeding fuel from a storage container (hopper) into a burn pot area, it produces a constant flame that requires little to no physical adjustments. Today's central heating systems operated with wood pellets as a renewable energy source can reach an efficiency factor of more than 90%.

Method

The pellet fuel is delivered from the storage facility or the day tank (single stoves) into the combustion chamber. With the heat generated, circuit water is heated in the pellet boiler. In central heating systems the hot water then runs through the heating circuit. The heat distribution is the same as other central heating systems. Unlike oil or gas heating, the inclusion of a hot water reservoir is recommended with pellet heating systems to save hot water until it is needed.

Benefits

Pellet-burning central heating system in basement of family home.

Most pellet stoves are self-igniting and cycle themselves on and off under thermostatic control. Stoves with automatic ignition can be equipped with remote controls. Recent innovations include integrated microcontroller monitoring of various safety conditions and can run diagnostic tests if an imminent problem arises.

A properly cleaned and maintained pellet stove should not create creosote, the sticky, flammable substance that causes chimney fires. Pellets burn very cleanly and create only a layer of fine fly ash as a byproduct of combustion. The grade of pellet fuel affects the performance and ash output. Premium-grade pellets produce less than one percent ash content, while standard or low grade pellets produce up to six percent ash. Pellet stove users should be aware of the extra maintenance required with a lower grade pellet, and that inconsistent wood quality can cause serious effects to the electronic machinery over a short period of time.

A pellet stove is normally associated with pelletized wood. However, many pellet stoves will also burn fuels such as grain, corn, seeds, or woodchips. In some pellet stoves, these fuels may need to be mixed with wood pellets. Pelletized trash (containing mostly waste paper) is also a fuel for pellet stoves.

Unlike wood stoves, which operate exclusively on a principle of chimney draft, a pellet stove must use specially sealed exhaust pipe to prevent exhaust gases escaping into the living space due to the air pressure produced by a combustion blower. Pellet stoves require certified double-walled venting, normally three or four inches in diameter with a stainless steel interior and galvanized exterior. Because pellet stoves have a forced exhaust system, they have the advantage of not always requiring a vertical rise to vent, although a 3-to-5-foot (0.91 to 1.52 m) vertical run to induce some draft is recommended to prevent leakage in the case of a power outage. Like a modern gas appliance, pellet stoves can be vented horizontally through an outside wall and terminated below the roof line, making it an excellent choice for structures without an existing chimney. If an existing chimney is available, manufacturers urge use of a correctly sized stainless steel liner the length of the chimney for proper drafting. Modern building techniques have created tightly sealed homes, forcing many pellet stove manufacturers to recommend their stoves be installed with outside air intake to ensure the stoves will run efficiently and prevent potential negative pressure within the home.

Pellet stoves are approved for use in mobile homes, while standard wood-burning stoves are not.

In many states pellet fuel is exempt from sales tax.

Principles of Operation

A pellet stove burn pot.

A pellet stove normally consists of these components, whether basic or complex:

- A hopper.
- An auger system.
- Two blower fans: combustion and convection.
- A firebox: burn pot and ash collection system, sometimes lined with ceramic fiber panels.

- Various safety features (vacuum switch, heat sensors).
- A controller.

To properly function, a pellet stove uses electricity and can be connected to a standard electrical outlet. A pellet stove, like an automatic coal stoker, is a consistent heater consuming fuel that is fed evenly from a refillable hopper into the burn pot (a perforated cast-iron or steel basin), through a motorized system. The most commonly used distributor is an auger system that consists of a spiral length of metal encased in a tube. This mechanism is either located above the burn pot or slightly beneath and guides a portion of pellet fuel from the hopper upwards until it falls into the burn pot for combustion.

Pellet stoves are routinely tested in laboratory for improved performances and different fuels.

Fan systems are necessary for clean, economical performance. The flame produced is concentrated and intense in the small area of the burn pot as a combustion blower introduces air into the bottom of the burn pot, while also forcing exhaust gases into the chimney. While some pellet stoves will be hot to the touch (especially on the viewing window), most manufacturers utilize a series of cast-iron or steel heat exchangers that run along the back and top areas of the visible firebox. With a convection blower, room air is circulated through the heat exchangers and directed into the living space. This method allows for a much higher efficiency than the radiant heat of a hand-fed wood or coal stove, and will in most cases cause the top, sides, and back of the stove to be at most warm to the touch. Along with convection air, an exhaust fan forces air from the firebox through special venting specifically made for pellet fuel. This cycle of circulation is an integral part of the combustion system as well, for the concentrated high temperature flame will quickly overheat the firebox. The possible problems associated with overheating are electrical component failure and flames traveling into the auger tube causing a hopper fire. As safeguards, all pellet stoves are equipped with heat sensors, and sometimes vacuum sensors, enabling the controller to shut down if an unsafe condition is detected. For daily maintenance, an ash vacuum is recommended. These are similar to shop vacs, but are designed for the removal of ash materials. These vacuums are available with a pellet stove kit which enables the cleaning of the interior areas of the stove which improves efficiency.

Pellet stoves can either be lit manually or through an automatic igniter. The igniter piece resembles a car's electric cigarette lighter heating coil. Most models have automatic ignition and can be readily equipped with thermostats or remote controls.

Corn Stove

A corn stove is designed for whole-kernel shelled corn kernel combustion and is similar to a pellet stove. The chief difference between a pellet stove and a dedicated corn stove is the addition of metal stirring rod within the burnpot or an active ash removal system. These vary in design slightly, but usually consist of one long metal stalk with smaller rods welded at a perpendicular angle, in order to churn the burn pot as it spins. An active ash removal system consists of augers at the bottom of the burn pot that evacuate the ash and clinkers. During a normal burn cycle, the sugar content within corn (and other similar bio-fuels) will cause the ashes to stick together, forming a hard mass. The metal stirring rod breaks apart these masses, causing a much more consistent burn. While there is demand to create stoves that are able to burn multiple fuels with minimal adjustments, some pellet stoves are not designed to stir fuel and cannot burn corn fuel.

BIOMASS ALLOCATION

Biomass allocation is a concept in plant biology which indicates the relative proportion of plant biomass present in the different organs of a plant. It can also be used for whole plant communities.

Rationale

Different organs of plants serve different functions. Leaves generally intercept light and fix carbon, roots take up water and nutrients, and stems and petioles display the leaves in a favourable position and transport various compounds within the plant. Depending on environmental conditions, plants may change their investment scheme, to make plants with relatively bigger root systems, or more leaves. This balance has been suggested to be a 'functional equilibrium', with plants that experience low water or nutrient supply investing more in roots, and plants growing under low light or CO_2 conditions investing more in leaves or stems. Alternatively, it is also known as the 'balanced growth hypothesis', or the 'optimal partitioning theory'. Next to environmentally-induced changes, there are also inherent differences in biomass allocation between species, and changes that depend on the age or size of plants.

Related Concepts

Biomass allocation is the end result of a number of processes which take place in the plant. It starts with the way how sugars are allocated over the different organs after having been fixed by the leaves in the process of photosynthesis (sugar allocation). Conceptually this is simple to envisage, but to quantify the flow of sugars is challenging and requires sophisticated machinery. For plants growing under steady state conditions, it is feasible to determine sugar-allocation by constructing a C-budget. This requires determination of the C-uptake by the whole plant during photosynthesis, and the C-losses of shoots and roots during respiration. Further C-losses may occur when sugars and other C-based compounds are exuded by the roots, or disappear as volatiles in the leaves. When these measurements are combined with growth measurements and the C-concentrations present in the biomass of leaves, stems and roots, C-budgets can be constructed from which sugar allocation is derived.

These C-budgets are instructive, but require extensive measurements. A next level of analysis is to measure the growth allocation: what is the increase in total biomass of a plant, and to what extent is the increase due to growth of leaves, of stems and of roots. In young plants, growth allocation is often quite similar to the actual biomass allocation. But especially in trees, there may be a high yearly turnover in leaves and fine roots, and a low turnover in stems, branches and thick roots. In those cases, the allocation of growth and the final biomass allocation may diverge quite strongly over the years.

There have been attempts to give these three different levels of allocation different names (a.o. partitioning, distribution, fractionation), but so far they have been applied inconsistently. The fractions of biomass present in leaves and roots are also relevant variables in Plant growth analysis.

Calculation and Units

A common way to characterize the biomass allocation of a vegetative plant is to separate the plant in the organs of interest (e.g. leaves, stems, roots) and determine the biomass of these organs – generally on a dry mass basis - independently. The Leaf Mass Fraction (LMF) is then calculated as leaf dry mass / total plant dry mass, the Stem Mass Fraction (SMF) as stem dry mass / total plant dry mass, and Root Mass Fraction (RMF) as root dry mass / total plant dry mass. Generally, units are $g\ g^{-1}$ (g organ / g total plant biomass).

For generative plants, there is the additional compartment related to reproduction (flowers and flower stalks, seeds or fruits). The relative amount of biomass present in this compartment is often indicated as 'Reproductive Effort'. A related variable which is often used in agronomy is the 'Harvest index'. Because roots are seldomly harvested, the harvest index is the amount of marketable product (often the seeds), relative to the total above-ground biomass.

Alternative terminology that has been used are Leaf, Stem and Root Mass Ratios, or shoot:root or root:shoot ratios. The latter two convey less information, as they do not discriminate between leaves and stems.

Normal Ranges

Young herbaceous plants generally have LMF values in the range of 0.3 - 0.7 g g^{-1} (0.5 on average), SMF values ranging from 0.04 - 0.4 (0.2 on average), and RMF values between 0.1 and 0.5 (0.3 on average). Young tree seedlings have values in the same range. For older and bigger plants, the LMF decreases and SMF increases. For large trees (> 1000 kg) LMF is below 0.05, SMF around 0.8 and RMF around 0.2 g g^{-1}. It should be noted that at that stage most of the stem biomass consists of highly lignified material, which still may serve the important function of contributing to the support function of stems, but is physiologically not active anymore.

Environmental Effects

The effect of the environment generally is as expected from the 'functional equilibrium' concept: plants decrease LMF and increase RMF when grown at high light levels as compared to low light. At low nutrient levels they invest more in roots and less in leaves as compared to high nutrient supply. However, changes are often smaller at different water supply, and effects of CO_2 concentration, UV-B radiation, ozone and salinity on allocation are generally negligible. Plants growing at higher temperature mostly decrease RMF and increase LMF.

A point of attention in the analysis of mass fractions is whether or not to correct for differences in size, when comparing plants that have been treated differently, or in the comparison of species. The rationale behind this is that mass factions often change with plant size (and developmental phase), and different treatments may have caused growth differences as well. Thus, for an assessment of whether plants *actively* changed their allocation scheme, plants of similar size should be compared. If size corrections are required, one could do an allometric analysis. A simple alternative is to plot mass fractions against total plant mass.

Differences between Species

Species of different families may have different allocation patterns. For example, species belonging to the Solanaceae have high LMF values, whereas Fagaceae have low LMF values, even after size-corrections. Grasses generally have lower LMF values that herbaceous dicots, with a much higher proportion of their biomass present in roots. Large evergreen trees have a larger fraction of their biomass allocated to leaves (LMF ~0.04) than deciduous species (LMF ~0.01).

References

- Biorefinery: bioenergyconsult.com, Retrieved 25 August, 2019
- Diener, stefan; zurbrügg, christian; tockner, klement (2009-06-05). "conversion of organic material by black soldier fly larvae: establishing optimal feeding rates". Waste management & research. Sage publications. 27 (6): 603–610. Doi:10.1177/0734242x09103838. Issn 0734-242x
- Bioreactors-types-6-types-of-bioreactors-used-in-bioprocess-technology, bioprocess-technology, biotechnology: biologydiscussion.com, Retrieved 31 July, 2019
- Hartwich, jens; bölscher, jens; schulte, achim (2014-09-19). "impact of short-rotation coppice on water and land resources". Water international. 39 (6): 813–825. Doi:10.1080/02508060.2014.959870. Issn 0250-8060
- What-are-energy-crops: basmati.com, Retrieved 3 July, 2019
- Dimitriou, ioannis; mola-yudego, blas; aronsson, pär (2012). "impact of willow short rotation coppice on water quality". Bioenergy research. 5 (3): 537–545. Doi:10.1007/s12155-012-9211-5
- "Global status of beccs projects 2010". Biorecro ab, global ccs institute. 2010. Archived from the original on 2014-05-09. Retrieved 2011-12-09
- M. R. Schmer, k. P. Vogel, r. B. Mitchell, and r. K. Perrin net energy of cellulosic ethanol from switchgrass pnas published january 7, 2008, doi:10.1073/pnas.0704767105

Biomass Feedstocks

3 CHAPTER

Any renewable and biological material that can be used directly as a fuel or converted to another form of fuel is known as a feedstock. Biomass feedstocks are the plants and materials used to derive fuels like ethanol, butanol, biodiesel, etc. The topics elaborated in this chapter will help in gaining a better perspective about different types of biomass feedstocks such as lignocellulosic biomass, miscanthus giganteus, landfill gas and jatropha curcas.

FEEDSTOCKS

A feedstock is defined as any renewable, biological material that can be used directly as a fuel, or converted to another form of fuel or energy product. Biomass feedstocks are the plant and algal materials used to derive fuels like ethanol, butanol, biodiesel, and other hydrocarbon fuels. Examples of biomass feedstocks include corn starch, sugarcane juice, crop residues such as corn stover and sugarcane bagasse, purpose-grown grass crops, and woody plants.

Ensuring a sustainable supply of high-quality biomass feedstock requires research and development to streamline all elements of the biomass feedstock supply chain—from plant breeding and genomics, to crop production and harvesting practices, to biomass preprocessing, transport, and storage systems. Sustainable feedstock production includes all the steps required to produce biomass feedstocks to the point where it is ready to be collected or harvested from the field or forest. Feedstock Logistics encompasses all the unit operations necessary to harvest the biomass and move it from the field or forest to the throat of the conversion process at the biorefinery, while ensuring that the delivered feedstock meets biorefinery physical and chemical quality specifications.

WOODY BIOMASS-DERIVED FEEDSTOCKS

Logging Residues

Logging residues are the unused portions of trees cut during logging operations, that are often left on-site. Logging residues are obtained from standard forest harvests, thinning young tree stands, or from forest stands that have been damaged by pests, storms, disease, or fires. Thinning is a standard silvicultural practice that includes removing smaller trees to reduce tree stand density (or number of trees per acre) to alleviate competition between trees, and improve stand health. Logging residues and wood from thinnings typically have little or no value in locations where markets do not exist.

During a forest harvest, merchantable trees are brought to a logging deck or landing site in the field where the timber is sorted by product type and transported to a nearby mill where the highest market value is obtained. Meanwhile, logging residues can be chipped on site and loaded into a tractor trailer or chip van for transport to a bioenergy facility or delivered as is to be processed at a bioenergy facility. Logging residues are typically processed in the field one of two ways- as clean chips or hog fuel. Clean chips do not include bark and are cut into short, thin wafers. Chips are used as a raw material for the production of paper, fiberboard, biomass fuel, and other products. Hog fuel refers to woody biomass and wood waste that has been ground in a tub grinder to produce a coarse mixture of bark and chips that vary in size. Hog fuel is primarily used in boilers that are able to handle the inconsistent mixture of differently sized pieces for energy production.

Lumber Mill Residues

Lumber mill residues include sawdust, shavings, and other waste from processing timber. This by-product is often used to fire the dry kilns or wood-fired boilers of the facility, thus producing some of the energy needed on site. Other times, residues are sold to other mills to make animal bedding, or pelletized and bagged for home heating in stoves. Throughout the Southeast, most mill residues are already utilized providing electricity to mill operations or sold to another mill to be converted to a higher-end product.

Other Wood Wastes–Yard Waste, Construction and Demolition Wood Waste

Some states or cities ban yard waste, or land clearing debris from traditional landfills and require that the material be managed at separate facilities. Wood waste facilities primarily process the collected material into mulch, compost, or hog fuel. Hog fuel is primarily used for energy in wood-fired boilers while most yard waste residues are processed into mulch by public or private entities and sold or given to the public. Wood waste from these sources may be problematic for generating electricity since the material often has trash such as fencing, bottles, nails, etc. that requires additional processing before supplying a power plant. Construction and demolition facilities accept wood recovered from construction and demolition projects, which are then converted into hog fuel or mulch.

Short-Rotation Woody Crops

The practice of using short-rotation woody crops (SRWC) or fast-growing trees for bioenergy is under extensive research and development. Cellulosic feedstocks such as

SRWC contain cellulose which is a carbohydrate and the principal component of the cell walls of trees and other higher-order plants, and are more difficult to break down than starchy grains such as corn. These cellulosic feedstocks are intended to be grown on marginal and other lands not commonly used for food production which helps to avoid the "food vs fuel" controversy. Research towards this end includes improving tree genetics and silvicultural practices, determining feedstock yield per acre on different soil productivity sites, and best ways to produce biofuels from the various SRWC.

An example of a wood waste pile.

SRWC species under research in the southeast United States include: hybrid poplars (populous spp.), sweetgum (Liquidambar styraciflua), American sycamore (Platanus occidentalis), loblolly pine (Pinus taeda), and eucalyptus (benthamii, grandis, urophylla). The hybrid poplars (sweetgum, American sycamore and loblolly pine) are able to be grown throughout most of the southeast. Eucalyptus species selected for biomass production in the southeast will be dependent on climate characteristics8 with Eucalyptus benthamii more apt to thrive in temperate climates and grandis, urophylla and their hybrids preferring more subtropical climates. SRWC are generally harvested in 1- to 6-year intervals, and coppiced woody crops can provide 3-6 harvests before replanting. Coppicing refers to cutting trees to the ground and then letting the shoots regrow from the main stump. SRWC can be harvested year round and is easier to transport, handle, and store than bioenergy grasses because of reduced storage and inventory costs and minimal loss or degradation of the product.

A research plot of hybrid poplars short rotation woody crops at the Williamsdale Biofuels Field Laboratory, Wallace, NC.

CROP RESIDUE

There are two types of agricultural crop residues. Field residues are materials left in an agricultural field or orchard after the crop has been harvested. These residues include stalks and stubble (stems), leaves, and seed pods. The residue can be ploughed directly into the ground, or burned first. In contrast, no-till, strip-till or reduced till agriculture practices are carried out to maximize crop residue cover. Good management of field residues can increase efficiency of irrigation and control of erosion. Simple line transect measurements can be used to estimate residue coverage. Process residues are materials left after the crop is processed into a usable resource. These residues include husks, seeds, bagasse, molasses and roots. They can be used as animal fodder and soil amendment, fertilizers and in manufacturing.

Economic Value

Most discussions about the economic value of crop residues focus on the equivalent fertilizer cost of the nutrients within. Although crop residues contain both macro- and micro nutrients, only values for the macro nutrients nitrogen, phosphorus, potassium and sulfur are economically significant.

Biofuel Production from Crop Residues

Because of the high carbohydrate content, crop residues can be considered as an appropriate feedstock to produce biofuels. Some algorithms have been developed to estimate the potential capacity of biofuel production from agricultural residues. Based on the experimental data obtained from a study that used ethanol organosolv pretreated rice straw to produce biohydrogen using *Enterobacter aerogenes*, the annual global amount of collectable rice straw (not total produced straw) for biofuel production was estimated about 249 million tonnes, that could approximately produce 355.78 kilotonnes of hydrogen and 11.32 million tonnes of lignin by the proposed organosolv technology and it was found that China contributes to about 32% of global potential capacity to produce biohydrogen from rice straw.

Mineralization

Nutrients in most crop residue are not immediately available for crop use. Their release (called mineralization) occurs over a period of years. The biological processes involved in soil nutrient cycles are complex. As a rough guide, cereal straw releases about 10 to 15 per cent of its nutrients and pea residues release about 35 percent of their nutrients by the next year.

The speed of mineralization depends on the nitrogen and lignin (fiber) content, soil moisture, temperature, and degree of mixing with the soil. N is released fairly quickly from residue when the content is higher than 1.5 per cent (such as in pea residues). In contrast, below 1.2 per cent (such as cereal residue), soil-available N is fixed (called immobilization) by the microbes as they decompose the residue.

Thus pea residue would have short- and long-term benefits to soil fertility, whereas cereal straw would reduce next year's soil supply of available nutrients. Over time, the nutrients fixed by soil microbes and humus are released and available to crops. Nutrients from residue are not fully recovered by crops. Just like fertilizer nutrients, nutrients released from crop residue into the soil are susceptible to losses such as leaching (N and S), denitrification (N), immobilization (N, P, K and S), and fixation (P and K).

Efficiency of Nutrient Uptake

The efficiency of nutrient uptake by crops from fertilizers or residue release is generally thought to be similar. For example, about 50 percent recovery of N in the above-ground plant in the first year. There is some residual benefit of fertilizers as the crops take up a small amount of the nutrients two and three years later. Fertilizer placement can significantly affect the efficiency of crop uptake. The impact of residue placement (buried by tillage or left on the surface in zero tillage) on nutrient cycling and efficiency is under study.

Thus, the practice of calculating the fertilizer equivalent value of the nutrients in crop residue is a reasonable guide to estimating the partial value of crop residues.

JATROPHA CURCAS

Jatropha curcas is a species of flowering plant in the spurge family, Euphorbiaceae, that is native to the American tropics, most likely Mexico and Central America. It is originally native to the tropical areas of the Americas from Mexico to Argentina, and has been spread throughout the world in tropical and subtropical regions around the world, becoming naturalized or invasive in many areas. The specific epithet, *"curcas"*, was first used by Portuguese doctor Garcia de Orta more than 400 years ago. Common names in English include physic nut, Barbados nut, poison nut, bubble bush or purging nut. In parts of Africa and areas in Asia such as India it is often known as "castor oil plant" or "hedge castor oil plant", but it is not the same as the usual castor oil plant, *Ricinus communis* (they are in the same family but different subfamilies).

J. curcas is a semi-evergreen shrub or small tree, reaching a height of 6 m (20 ft) or more. It is resistant to a high degree of aridity, allowing it to grow in deserts. It contains phorbol esters, which are considered toxic. However, edible (non-toxic) provenances native to Mexico also exist, known by the local population as piñón manso, xuta, chuta, aishte, among others. *J. curcas* also contains compounds such as trypsin inhibitors, phytate, saponins and a type of lectin known as curcin.

The seeds contain 27–40% oil (average: 34.4%) that can be processed to produce a high-quality biodiesel fuel, usable in a standard diesel engine. Edible (non-toxic) provenances can be used for animal feed and food.

Botanical Features

- Leaves: The leaves have significant variability in their morphology. In general, the leaves are green to pale green, alternate to subopposite, and three- to five-lobed with a spiral phyllotaxis.
- Flowers: Male and female flowers are produced on the same inflorescence, averaging 20 male flowers to each female flower, or 10 male flowers to each female flower. The inflorescence can be formed in the leaf axil. Plants occasionally present hermaphroditic flowers.
- Fruits: Fruits are produced in winter, or there may be several crops during the year if soil moisture is good and temperatures are sufficiently high. Most fruit production is concentrated from midsummer to late fall with variations in production peaks where some plants have two or three harvests and some produce continuously through the season.
- Seeds: The seeds are mature when the capsule changes from green to yellow. The seeds contain around 20% saturated fatty acids and 80% unsaturated fatty acids, and they yield 25–40% oil by weight. In addition, the seeds contain other chemical compounds, such as saccharose, raffinose, stachyose, glucose, fructose, galactose, and protein. The oil is largely made up of oleic and linoleic acids. Furthermore, the plant also contains curcasin, arachidic, myristic, palmitic, and stearic acids and curcin.
- Genome: The whole genome was sequenced by *Kazusa DNA Research Institute*, Chiba Japan in October 2010.

Cultivation

Cultivation is uncomplicated. *Jatropha curcas* grows in tropical and subtropical regions. The plant can grow in wastelands and grows on almost any terrain, even on gravelly, sandy and saline soils. It can thrive in poor and stony soils, although new research suggests that the plant's ability to adapt to these poor soils is not as extensive as had been previously stated. Complete germination is achieved within 9 days. Adding manure during the germination has negative effects during that phase, but is favorable if applied after germination is achieved. It can be propagated by cuttings, which yields faster results than multiplication by seeds.

The flowers only develop terminally (at the end of a stem), so a good ramification (plants presenting many branches) produces the greatest amount of fruits. The plants are self-compatible. Another productivity factor is the ratio between female and male flowers within an inflorescence, more female flowers mean more fruits. *Jatropha curcas* thrives on a mere 250 mm (10 in) of rain a year, and only during its first two years does it need to be watered in the closing days of the dry season. Ploughing and planting are not needed regularly, as this shrub has a life expectancy of approximately forty years. The use of pesticides is not necessary, due to the pesticidal and fungicidal properties of the plant. It is used in rural Bengal for dhobi itch (a common fungal infection of the skin).

Jatropha curcas seeds.

While *Jatropha curcas* starts yielding from 9–12 months time, the best yields are obtained only after 2–3 years time. The seed production is around 3.5 tons per hectare (seed production ranges from about 0.4 t/ha in the first year to over 5 t/ha after 3 years). If planted in hedges, the reported productivity of *Jatropha* is from 0.8 to 1.0 kg of seed per meter of live fence.

Propagation

Jatropha curcas can easily be propagated by both seed or cuttings. Some people recommend propagation by seed for establishment of long-lived plantations. When jatropha plants develop from cuttings, they produce many branches but yield fewer seeds and do not have enough time to develop their taproot, which makes them sensitive to wind erosion. The seeds exhibit orthodox storage behaviour and under normal treatment and storage will maintain viability at high percentages for eight months to a year. Propagation through seed (sexual propagation) leads to a lot of genetic variability in terms of growth, biomass, seed yield and oil content. Clonal techniques can help in overcoming these problems. Vegetative propagation has been achieved by stem cuttings, grafting, budding as well as by air layering techniques. Cuttings should be taken preferably from juvenile plants and treated with 200 micro gram per litre of IBA (rooting hormone) to ensure the highest level of rooting in stem cuttings. Cuttings strike root easily stuck in the ground without use of hormones.

Propagation of Jatropha curcas by stem cutting.

Propagation of jatropha curcas by grafting.

Propagation of jatropha curcas by air layering.

Jatropha crude oil.

Processing

Seed extraction and processing generally needs specialized facilities.

Oil content varies from 28% to 30% and 80% extraction, one hectare of plantation will give 400 to 600 litres of oil if the soil is average.

The oily seeds are processed into oil, which may be used directly ("Straight Vegetable Oil") to fuel combustion engines or may be subjected to transesterification to produce biodiesel. Jatropha oil is not suitable for human consumption, as it induces strong vomiting and diarrhea.

Biofuel

When jatropha seeds are crushed, the resulting jatropha oil can be processed to produce a high-quality biofuel or biodiesel that can be used in a standard diesel car or further processed into jet fuel, while the residue (press cake) can also be used as biomass feedstock to power electricity plants, used as fertilizer (it contains nitrogen, phosphorus and potassium). The cake can also be used as feed in digesters and gasifiers to produce biogas.

Jatropha plantation in the dry center/west of the Paraguay Chaco.

There are several forms of biofuel, often manufactured using sedimentation, centrifugation, and filtration. The fats and oils are turned into esters while separating the glycerin. At the end of the process, the glycerin settles and the biofuel floats. The process through which the glycerin is separated from the biodiesel is known as transesterification. Glycerin is another by-product from Jatropha oil processing that can add value to the crop. Transesterification is a simple chemical reaction that neutralizes the free fatty acids present in any fatty substances in Jatropha. A chemical exchange takes place between the alkoxy groups of an ester compound by an alcohol. Usually, methanol and ethanol are used for the purpose. The reaction occurs by the presence of a catalyst, usually sodium hydroxide (NaOH) or caustic soda and potassium hydroxide (KOH), which forms fatty esters (e.g., methyl or ethyl esters), commonly known as biodiesel. It takes approximately 10% of methyl alcohol by weight of the fatty substance to start the transesterification process.

"From planting to picking. Treat your jatropha plant as well as possible to make the yield as large as possible."

Estimates of *Jatropha* seed yield vary widely, due to a lack of research data, the genetic diversity of the crop, the range of environments in which it is grown, and *Jatropha*'s perennial life cycle. Seed yields under cultivation can range from 1,500 to 2,000 kilograms per hectare, corresponding to extractable oil yields of 540 to 680 litres per hectare (58 to 73 US gallons per acre). In 2009 *Time* magazine cited the potential for as much as 1,600 gallons of diesel fuel per acre per year. The plant may yield more than four times as much fuel per hectare as soybean, and more than ten times that of maize (corn), but at the same time it requires five times as much water per unit of energy produced as does corn. A hectare of jatropha has been claimed to produce 1,892 litres of fuel. However, as it has not yet been domesticated or improved by plant breeders, yields are variable.

Jatropha can also be intercropped with other cash crops such as coffee, sugar, fruits and vegetables.

In 2007 Goldman Sachs cited *Jatropha curcas* as one of the best candidates for future biodiesel production. However, despite its abundance and use as an oil and reclamation plant, none of the *Jatropha* species has been properly domesticated and, as a result, its productivity is variable, and the long-term impact of its large-scale use on soil quality and the environment is unknown.

In 2008 researchers at Daimler Chrysler Research explored the use of jatropha oil for automotive use, concluding that although jatropha oil as fuel "has not yet reached optimal quality, it already fulfills the EU norm for biodiesel quality". Archer Daniels Midland Company, Bayer CropScience and Daimler AG have a joint project to develop jatropha as a biofuel. Three Mercedes cars powered by Jatropha diesel have already put some 30,000 kilometres behind them. The project is supported by DaimlerChrysler and by the German Association for Investment and Development (Deutschen Investitions- und Entwicklungsgesellschaft, DEG).

Jet Fuel

Aviation fuels may be more widely replaced by biofuels such as jatropha oil than fuels for other forms of transportation. There are fewer planes than cars or trucks and far fewer jet fueling stations to convert than gas stations. To fulfil the yearly demand for aviation fuel, based on demand in 2008 (fuel use has since grown), an area of farmland twice the size of France would need to be planted with jatropha, based on average yields of mature plantations on reasonably good, irrigated land.

On December 30, 2008, Air New Zealand flew the first successful test flight from Auckland with a Boeing 747 running one of its four Rolls-Royce engines on a 50:50 blend of jatropha oil and jet A-1 fuel. In the same press release, Air New Zealand announced plans to use the new fuel for 10% of its needs by 2013. At the time of this test, jatropha oil was much cheaper than crude oil, costing an estimated $43 a barrel or about one-third of the June 4, 2008 closing price of $122.30 for a barrel of crude oil.

On January 7, 2009 Continental Airlines successfully completed a test flight from Houston, Texas using a 50/50 mixture of algae/jatropha-oil-derived biofuel and Jet A in one of the two CFM56 engines of a Boeing 737-800 Next Generation jet. The two-hour test flight could mark another promising step for the airline industry to find cheaper and more environmentally friendly alternatives to fossil fuel.

On April 1, 2011 Interjet completed the first Mexican aviation biofuels test flight on an Airbus A320. The fuel was a 70:30 traditional jet fuel biojet blend produced from Jatropha oil provided by three Mexican producers, Global Energías Renovables (a wholly owned subsidiary of U.S.-based Global Clean Energy Holdings), Bencafser S.A. and Energy JH S.A. Honeywell's UOP processed the oil into Bio-SPK (Synthetic Paraffinic Kerosene). Global Energías Renovables operates the largest Jatropha farm in the Americas.

On October 28, 2011 Air China completed the first successful demonstration flight by a Chinese airline that used jatropha-based biofuel. The mixture was a 50:50 mix of conventional jet fuel blended with jatropha oil from China National Petroleum Corp. The 747-400 powered one of its four engines on the fuel mixture during the 1-hour flight around Beijing airport.

On August 27, 2018 SpiceJet completed the first successful test flight by an Indian airline which used jatorpha based biofuel. The ratio of conventional jet fuel to jatropha oil was 25:75.

Carbon Dioxide Sequestration

According to a 2013 study published by the European Geosciences Union, the jatropha tree may have applications in the absorption of carbon dioxide, whose sequestration is important in combating climate change. This small tree is very resistant to aridity so it can be planted in hot and dry land in soil unsuitable for food production. The plant does need water to grow though, so coastal areas where desalinated seawater can be made available are ideal.

Use in Developing World

Currently the oil from *Jatropha curcas* seeds is used for making biodiesel fuel in Philippines, Pakistan and in Brazil, where it grows naturally and in plantations in the southeast, north, and northeast of Brazil. Likewise, jatropha oil is being promoted as an easily grown biofuel crop in hundreds of projects throughout India and other developing countries. Large plantings and nurseries have been undertaken in India by many research institutions, and by women's self-help groups who use a system of microcredit to ease poverty among semiliterate Indian women. The railway line between Mumbai and Delhi is planted with *jatropha* and the train itself runs on 15–20% biodiesel. In Africa, cultivation of *jatropha* is being promoted and it is grown successfully in countries such as Mali. In the Gran Chaco of Paraguay, where a native variety (*Jatropha matacensis*) also grows, studies have shown the suitability of Jatropha cultivation and agro producers are starting to consider planting in the region.

Myanmar

Myanmar is also actively pursuing the use of jatropha oil. On 15 December 2005, then-head of state, Senior General Than Shwe, said "the States and Divisions concerned are to put 50,000 acres (200 km^2) under the physic nut plants [Jatropha] each within three years totalling 700,000 acres (2,800 km^2) during the period". On the occasion of Burma's Peasant Day 2006, Than Shwe described in his a message that "For energy sector which is an essential role in transforming industrial agriculture system, the Government is encouraging for cultivation of physic nut plants nationwide and the technical know how that can refine physic nuts to biodiesel has also identified." He would like to urge peasants to cultivate physic nut plants on a commercial scale with major aims for emergence of industrial agriculture system, for fulfilling rural electricity supply and energy needs, for supporting rural areas development and import substitute economy.

In 2006, the chief research officer at state-run Myanma Oil and Gas Enterprise said Burma hoped to completely replace the country's oil imports of 40,000 barrels a day with home-brewed, jatropha-derived biofuel. Other government officials declared Burma would soon start exporting jatropha oil. Despite the military's efforts, the jatropha campaign apparently has largely flopped in its goal of making Burma self-sufficient in fuel.

Z.G.S. Bioenergy has started Jatropha Plantation Projects in Northern Shan State, the company has begun planting Jatropha plants during late June 2007 and will start producing seeds by 2010.

As of 2011 scepticism about the "miracle" properties of *Jatropha* has been voiced. For example: "The idea that jatropha can be grown on marginal land is a red herring", according to Harry Stourton, former business development director of UK-based Sun Biofuels, which attempted to cultivate *Jatropha* in Mozambique and Tanzania. "It does grow on marginal land, but if you use marginal land you'll get marginal yields," he said. Sun Biofuels, after failing to adequately compensate local farmers for the land acquired for their plantation in Tanzania, pay workers severance, or deliver promised supplies to local villagers, went bankrupt later in 2011, the villager farmland being sold to an offshore investment fund.

An August 2010 article warned about the actual utility and potential dangers of reliance on *Jatropha* in Kenya. Major concerns included its invasiveness, which could disrupt local biodiversity, as well as damage to water catchment areas.

Jatropha curcas is lauded as being sustainable, and that its production would not compete with food production, but the jatropha plant needs water like every other crop to grow. This could create competition for water between the jatropha and other edible food crops. In fact, jatropha requires five times more water per unit of energy than sugarcane and corn.

LIGNOCELLULOSIC BIOMASS

Lignocellulose is a generic term for describing the main constituents in most plants, namely cellulose, hemicelluloses, and lignin. Lignocellulose is a complex matrix, comprising many different polysaccharides, phenolic polymers and proteins. Cellulose, the major component of cell walls of land plants, is a glucan polysaccharide containing large reservoirs of energy that provide real potential for conversion into biofuels. Lignocellulosic biomass consists of a variety of materials with distinctive physical and chemical characteristics. It is the non-starch based fibrous part of plant material.

First-generation biofuels (produced primarily from food crops such as grains, sugar beet and oil seeds) are limited in their ability to achieve targets for oil-product substitution, climate change mitigation, and economic growth. Their sustainable production is under scanner, as is the possibility of creating undue competition for land and water used for food and fibre production.

The cumulative impacts of these concerns have increased the interest in developing biofuels produced from non-food biomass. Feedstocks from ligno-cellulosic materials include cereal straw, bagasse, forest residues, and purpose-grown energy crops such as vegetative grasses and short rotation forests. These second-generation biofuels could avoid many of the concerns facing first-generation biofuels and potentially offer greater cost reduction potential in the longer term.

The largest potential feedstock for ethanol is lignocellulosic biomass, which includes materials such as agricultural residues (corn stover, crop straws and bagasse), herbaceous crops (alfalfa, switchgrass), short rotation woody crops, forestry residues, waste paper and other wastes (municipal and industrial). Bioethanol production from these feedstocks could be an attractive alternative for disposal of these residues. Importantly, lignocellulosic feedstocks do not interfere with food security. Moreover, bioethanol is very important for both rural and urban areas in terms of energy security reason, environmental concern, employment opportunities, agricultural development, foreign exchange saving, socioeconomic issues etc.

Lignocellulosic biomass consists mainly of lignin and the polysaccharides cellulose and hemicellulose. Compared with the production of ethanol from first-generation feedstocks, the use of lignocellulosic biomass is more complicated because the polysaccharides are more stable and the pentose sugars are not readily fermentable by Saccharomyces cerevisiae. In order to convert lignocellulosic biomass to biofuels the polysaccharides must first be hydrolysed, or broken down, into simple sugars using either acid or enzymes. Several biotechnology-based approaches are being used to overcome such problems, including the development of strains of Saccharomyces cerevisiae that can ferment pentose sugars, the use of alternative yeast species that naturally ferment pentose sugars, and the engineering of enzymes that are able to break down cellulose and hemicellulose into simple sugars.

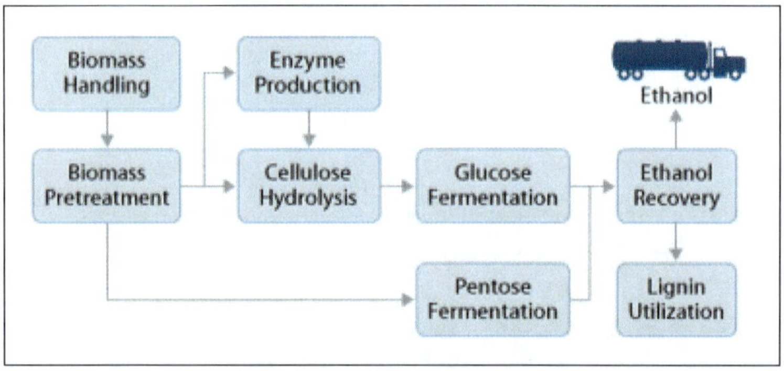

Schematic of a biochemical cellulosic ethanol production process.

Lignocellulosic processing pilot plants have been established in the EU, in Denmark, Spain and Sweden. The world's largest demonstration facility of lignocellulose ethanol (from wheat, barley straw and corn stover), with a capacity of 2.5 Ml, was first established by Iogen Corporation in Ottawa, Canada. Many other processing facilities are now in operation or planning throughout the world.

Economically, lignocellulosic biomass has an advantage over other agriculturally important biofuels feedstocks such as corn starch, soybeans, and sugar cane, because it can be produced quickly and at significantly lower cost than food crops. Lignocellulosic biomass is an important component of the major food crops; it is the non-edible portion of the plant, which is currently underutilized, but could be used for biofuel production. In short, lignocellulosic biomass holds the key to supplying society's basic needs for sustainable production of liquid transportation fuels without impacting the nation's food supply.

MISCANTHUS GIGANTEUS

Miscanthus x giganteus (miscanthus giganteus, giant miscanthus, elephant grass) is a sterile hybrid of Miscanthus sinensis and Miscanthus sacchariflorus. It can grow to heights of more than 4 metres (13 ft) in one growing season (from the third season onwards). Just like Pennisetum purpureum and Saccharum ravennae it is also called elephant grass.

M. x giganteus' perennial nature, its ability to grow on marginal land, its water efficiency, non-invasiveness, low fertilizer needs, significant carbon sequestration and high yield have sparked a lot of interest among researchers, with some arguing that it has "ideal" energy crop properties. Some argue that it has the potential to be a greenhouse gas (GHG) negative fuel, while others highlight its water cleaning and soil enhancing qualities. There are practical and economic challenges related to its use in the existing, fossil based combustion infrastructure, however. Torrefaction and other fuel upgrading techniques are being explored as countermeasures to this problem.

Use Areas

M. x giganteus is mainly used as raw material for solid biofuels. It can be burned directly, or processed further into pellets or briquettes. It can also be used as raw material for liquid biofuels or biogas.

It is possible to use Miscanthus as a building material, and as insulation. Materials produced from Miscanthus include fiberboards, composite Miscanthus/wood particleboards, and blocks. It can be used as raw material for pulp and fibers as well as molded products such as eco-friendly disposable plates, cups, cartons,etc. Miscanthus has a pulp yield of 70-80% compared to dry weight, due to the high holocellulose content. The pulp can be processed further into methylcellulose and used as a food additive and in many industrial applications. Miscanthus fiber provides raw material for reinforcement of biocomposite or synthetic materials. In agriculture, Miscanthus straw is used in soil mulching to retain soil moisture, inhibit weed growth, and prevent erosion. Further, Miscanthus' high carbon to nitrogen ratio makes it inhospitable to many microbes, creating a clean bedding for poultry, cattle, pigs, horses, and companion animals. Miscanthus used as horse bedding can be combined with making organic fertilizer. Miscanthus can be used as a healthy fiber source in pet food.

Life Cycle

Propagation

Miscanthus x giganteus is propagated by cutting the rhizomes (its underground stems) into small pieces, and then re-planting those pieces 10 cm (4 in) underground. One hectare (2.5 acres) of Miscanthus rhizomes, cut into pieces, can be used to plant 10-30 hectares of new Miscanthus fields (multiplication factor 10-30). Rhizome propagation is a labor-intensive way of planting new crops, but only happens once during a crop's lifetime. New and cheaper propagation techniques is underway, which seem to increase the multiplication factor from 10-30 to 1000-2000. A halving of the cost is predicted.

Management

A limited amount of herbicide should only be applied at the beginning of the first two seasons; after the second year the dense canopy and the mulch formed by dead leaves effectively reduces weed growth. Other pesticides are not needed. Because of Miscanthus' high nitrogen efficiency, fertilizer is also usually not needed. Mulch film, on the other hand, helps both M. x giganteus and various seed based hybrids to grow faster and taller, with a larger number of stems per plant, effectively reducing the establishment phase from three years to two. The reason seems to be that this plastic film keeps the humidity in the topsoil and increases the temperature.

Yield

Young miscanthus test crop in England.

Miscanthus x giganteus is close to the theoretical maximum efficiency at turning solar radiation into biomass, and its water use efficiency is among the highest of any crop. It has twice the water use efficiency as its fellow C4 plant maize, and four times the efficiency as the C3 plant wheat. The efficiency has as an important consequence: It makes Miscanthus x giganteus fields energy dense. Since Miscanthus has an energy content of 18 GJ per dry tonne, the typical UK dry yield of 11-14 tonnes per hectare produce 200-250 gigajoules of energy per hectare per year (200-250 GJ ha-1 yr-1). This compares favorably to maize (98 GJ), oil seed rape (25 GJ), and wheat/sugar beet (7-15 GJ), underlining the differences between first and second generation bioenergy crops.

In Europe the peak (autumn) dry mass yield has been measured to 10–40 tonnes per hectare per year (4–16 tonnes per acre per year), depending on location, with a mean peak dry mass yield of 22 tonnes. European yields are highest in southern Europe. Trials in Illinois, USA, had yields 10–15 tonnes per acre (25–37 t/ha). Like Europe, yields decrease as you move north. M. x giganteus has been shown to yield two to three times more than switchgrass.

Peak yield is reached at the end of summer but harvest is typically delayed until winter or early spring. Yield is roughly 33% lower at this point because of leaves drop, but the combustion quality is higher. Delayed harvest also allows nitrogen to move back into the rhizome for use by the plant in the following growing season.

Yield - Arable Land

In Germany, Felten et al. did a 16-year trial on arable land and concluded with a mean winter/spring yield of 15 tonnes per hectare per year (6.1 tonnes per acre per year). McCalmont et al. estimates a mean UK yield of 10-15 tonnes (if harvested in the spring), while Hastings et al. estimates a "pessimistic" UK mean yield of 10.5 tonnes. Nsanganwimana et al. summarizes several trials, and give these numbers:

- Austria: Autumn harvest 17-30. Winter harvest 22.
- Denmark: Autumn harvest 17. Winter harvest 10.
- France: Autumn harvest 42-49. Winter harvest 30.
- Germany: Autumn harvest 17-30. Winter harvest 10-20.
- Portugal: Autumn harvest 39. Winter harvest 26-30.
- The Netherlands: Autumn harvest 25. Winter harvest 16-17.
- Spain: Winter harvest 14.
- UK: Winter harvest 11-17.

Planting and harvest.

Yield - Marginal Land

Marginal land is land with issues that limits growth, for instance low water and nutrient storage capacity, high salinity, toxic elements, poor texture, shallow soil depth, poor drainage, low fertility, or steep terrain. Depending on how the term is defined, between 1.1 and 6.7 billion hectares of marginal land exists in the world. For comparison, Europe consists of roughly 1 billion hectares (10 million km^2, or 3.9 million square miles), and Asia 4.5 billion hectares (45 million km^2, or 17 million square miles).

Quinn et al. identified Miscanthus x giganteus as a crop that is moderately or highly tolerant of multiple environmental stressors, specifically, heat, drought, flooding, salinity (below 100 mM), and cool temperatures (down to −3.4 °C, or 25 °F). This robustness makes it possible to establish relatively high-yielding Miscanthus fields on marginal land, Nsanganwimana et al. mentions wastelands, coastal areas, damp habitats, grasslands, abandoned milling sites, forest edges, streamsides, foothills and mountain slopes as viable locations. Likewise, Stavridou et al. concluded that 99% of Europe's saline, marginal lands can be used for M. x giganteus plantations, with only an expected maximum yield loss of 11%. Since salinity up to 200 mM does not affect roots and rhizomes, carbon sequestration carry on unaffected. Lewandowski et al. found a yield loss of 36% on a marginal site limited by low temperatures (Moscow), compared to maximum yield on arable land in central Europe. They also found a yield loss of 21% on a marginal site limited by drought (Turkey), compared to maximum yields on arable soil in central Europe.

Steep, marginal land.

Further, Nsanganwimana et al. found that M. x giganteus grows well in soils contaminated by metals, or by industrial activities in general. For instance, in one trial, it was found that M. x giganteus absorbed 52% of the lead content and 19% of the arsenic content in the soil after three months. The absorption stabilizes the pollutants so they don't travel into the air (as dust), into ground water, neighbouring surface waters, or neighbouring areas used for food production. If contaminated Miscanthus is used as fuel, the combustion site need to install the appropriate equipment to handle this situation. On the whole though, Miscanthus is a suitable crop for combining biomass production and ecological restoration of contaminated and marginal land." Clifton-Brown et al. concludes that Miscanthus x giganteus can contribute to the sustainable intensification of agriculture, allowing farmers to diversify and provide biomass for an expanding market without compromising food security."

Carbon Sequestration

Soil Carbon Input/Output

Plants sequester carbon through photosynthesis, by exchanging O_2 (oxygen) for CO_2, thus keeping the carbon (C) to itself. When the plants move carbon to the roots, it does not stay down there forever however; soil carbon is a balance between the decay of the initial soil carbon and the rate of input Plant derived soil carbon is a continuum, ranging from living biomass to humus, and it decays in different stages, ranging from months (decomposable plant material; DPM) to hundreds of years (humus). The rate of decay depends on many factors, for instance plant species, soil, temperature and humidity, but as long as fresh new carbon is inputted, a certain amount of carbon stay in the ground – in fact Poeplau et al. did not find any indication of decreasing SOC [soil organic carbon] accumulation with age of the plantation indicating no SOC saturation within 15–20 years. The amount of carbon in the ground under Miscanthus fields is thus seen to increase during the entire life of the crop, albeit with a slow start because of the initial tilling (plowing, digging) and the relatively low amounts of carbon input in the establishment phase. (Tilling induces soil aeration, which accelerates the soil carbon decomposition rate, by stimulating soil microbe populations. Also, tilling makes it easier for the oxygen (O) atoms in the atmosphere to attach to carbon (C) atoms in the soil, producing CO_2). Felten et al. argues that high proportions of pre- and direct-harvest residues (e.g. dead leaves), direct humus accumulation, the well-developed and deep-reaching root system, the low decomposition rates of plant residues due to a high C : N ratio (carbon to nitrogen ratio), and the absence of tillage and subsequently less soil aeration are the reasons for the high carbon sequestration rates.

At the end of each season, the plant pulls the nutrients to the ground. The color shifts from green to yellow/brown.

Net Annual Carbon Accumulation

A number of studies try to quantify the net amount of below-ground carbon accumulation each year, after decay is accounted for, in various locations and under various circumstances.

Dondini et al. found 32 tonnes more carbon per hectare (13 tonnes per acre) under a 14 year old Miscanthus field than in the control site, suggesting a combined (C3 plus C4) accumulation rate of 2.29 tonnes per hectare (0.91 long ton/acre), or 38% of total harvested carbon per year. Likewise, Milner et al. suggest a mean carbon accumulation rate for the whole of the UK of 2.28 tonnes per hectare (0.91 long ton/acre) per year (also 38% of total harvested carbon per year), given that some unprofitable land (0.4% of total) is excluded. Nakajima et al. found an accumulation rate of 1.96 (± 0.82) tonnes per hectare per year below a university test site in Sapporo, Japan (0.79 per acre), equivalent to 16% of total harvested carbon per year. The test was shorter though, only 6 years. Hansen et al. found an accumulation rate of 0.97 tonne per hectare per year (0.39 tonnes per acre per year) over 16 years under a test site in Hornum, Denmark, equivalent to 28% of total harvested carbon per year. McCalmont et al. compared a number of individual European reports, and found accumulation rates ranging from 0.42 to 3.8 tonnes per hectare per year, with a mean accumulation rate of 1.84 tonne (0.74 tonnes per acre per year), or 20% of total harvested carbon per year.

Transport and Combustion Challenges

Biomass in general, including Miscanthus x giganteus, have different properties compared to coal, for instance when it comes to handling and transport, grinding, and combustion. This makes sharing the same logistics, grinding and combustion infrastructure difficult. Often new biomass handling facilities have to be built instead, which increases cost. Together with the relatively high cost of feedstock, this often lead to the well-known situation where biomass projects has to receive subsidies to be economically viable. A number of fuel upgrading technologies are currently being explored however that makes biomass more compatible with the existing infrastructure. The most mature of these is torrefaction, basically an advanced roasting technique which – when combined with pelleting or briquetting – significantly influences both the handling and transport properties, grindability and combustion efficiency.

Energy Density and Transport Costs

Transport of bulky, water absorbing Miscanthus bales in England.

Miscanthus bales and chips have a bulk density of approximately 150 kg/m^3, while briquettes have a bulk density of up to 600 kg/m^3. Torrefaction increases bulk density further, as approximately two thirds of the original mass is retained as a solid product, while approximately one third of the original mass was converted to gas during the process. The finished, torrefied, solid product, in the form of pellets or briquettes, still contains approximately 85% of the original energy, however. Basically the mass part reduce more than the energy part, and the consequence is that the calorific value of torrefied biomass approaches that of medium grade coal - typically the calorific value increases from 18 GJ per tonne dry mass to 23 GJ per tonne torrefied mass.

The higher energy density means lower transport costs, and a decrease in transport-related GHG emittance. The IEA (International Energy Agency) has calculated energy and GHG costs for regular and torrefied pellets/briquettes. When making pellets and shipping them from Indonesia to Japan, a minimum 6.7% of energy savings or 14% GHG savings is expected when switching from regular to torrefied. This number increases to 10.3% energy savings and 33% GHG savings when making and shipping minimum 50mm briquettes instead of pellets. The longer the route, the bigger the savings. The relatively short supply route from Russia to the UK equals energy savings of 1.8%, while the longer supply route from southeast USA to the Amsterdam-Rotterdam-Antwerp (ARA) area is 7.1%. From southwest Canada to ARA 10.6%, southwest USA to Japan 11%, and Brazil to Japan 11.7% (all these savings are for pellets only.)

Water Absorption and Transport Costs

Torrefaction also converts the biomass from a hydrophilic (water absorbing) to a hydrophobic (water repelling) state. Water repelling briquettes can be transported and stored outside, which simplifies the logistics operation and decreases cost. Almost all biological activity is stopped, reducing the risk of fire and biological decomposition like rotting.

Uniformity and Customization

Generally, torrefaction is seen as a gateway for converting a range of very diverse feedstocks into a uniform and therefore easier to deal with fuel. The fuel's parameters can be changed to meet customers demands, for instance durability, water resistance, ash composition, torrefaction degree, geometrical form, and type of feedstock. The possibility to use different types of feedstock improves the fuel's availability and supply reliability.

Grindability

Coal grinders.

Unprocessed M. x giganteus has strong fibers, making grinding into equally sized, very small particles (below 75 µm / 0.075 mm) difficult to achieve. Coal chunks are typically ground to that size because such small, even particles combust stabler and more efficient than the larger fuel chunks. While coal has a score on the Hardgrove Grindability Index (HGI) of 30-100 (higher numbers means it is easier to grind), unprocessed Miscanthus has a score of 0. During torrefaction however, the hemi-cellulose fraction which is responsible for the fibrous nature of biomass is degraded, thereby improving its grindability." Bridgeman et al. measured a HGI of 79 for torrefied Miscanthus, Smith et al. measured a HGI of 150 for Miscanthus pre-treated with hydrothermal carbonisation: while the IEA estimates a HGI of 23-53 for torrefied biomass in general. UK coal scores between 40 and 60 on the HGI scale. The IEA estimates an 80-90% drop in energy use required to grind biomass that has been torrefied.

The relatively easy grinding of torrefied Miscanthus makes a cost-effective conversion to fine particles possible, which subsequently makes efficient combustion with a stable flame possible. Ndibe et al. found that the level of unburnt carbon decreased with the introduction of torrefied biomass", and that the torrefied biomass flames were stable during 50% cofiring and for the 100% case as a result of sufficient fuel particle fineness."

Chlorine and Corrosion

Raw miscanthus biomass has a relatively high chlorine amount, which is problematic in a combustion scenario because, as Ren et al. explains, the likelihood of corrosion depends significantly on the content of chlorine in the fuel Likewise, Johansen et al. states that the release of Cl-associated [chlorine-associated] species during combustion is the main cause of the induced active corrosion in the grate combustion of biomass." Chlorine in different forms, in particular combined with potassium as potassium chloride, condensates on relatively cooler surfaces inside the boiler and creates a corrosive deposit layer. The corrosion damages the boiler, and in addition the physical deposit layer itself reduce heat transfer efficiencey, most critically inside the heat exchange mechanism. Chlorine and potassium also lowers the ash melting point considerably compared to coal. Melted ash, known as slag or clinker, sticks to the bottom of the boiler, and increase maintenance costs.

In order to reduce chlorine (and moisture) content, M. x giganteus is usually harvested dry, in early spring, but this late harvest practice is still not enough of a countermeasure to achieve corrosion-free combustion.

However, Ren et al. found that 59.1 wt%, 60.7 wt% and 77.4 wt% of the chlorine contents of olive residues, DDGS and corn straw, respectively, were released during torrefaction", and concludes that chlorine emissions were drastically lower, by 2–5 times, than those of their raw biomass precursors." Chlorine release during the torrefaction process itself is more manageble than chlorine release during combustion, because the prevailing temperatures during the former process are below the melting and vaporization temperatures of the alkali salts of chlorine, thus minimizing their risks of slagging, fouling and corrosion in furnaces."

For potassium, Kambo et al. found a 30% reduction for torrefied miscanthus, while Ren et al. found an 86% increase for torrefied corn stover. However, potassium is dependent on chlorine to form potassium chloride; with a low level of chlorine, the potassium chloride deposits reduce proportionally.

Li et al. concludes that the process of torrefaction transforms the chemical and physical properties of raw biomass into those similar to coal, which enables utilization with high substitution ratios of biomass in existing coal-fired boilers without any major modifications." Likewise, Bridgeman et al. states that since torrefaction removes moisture, creates a grindable, hydrophobic and solid product with an increased energy density, torrefied fuel no longer requires separate handling facilities when co-fired with coal in existing power stations." Smith et al. makes a similar point in regard to hydrothermal carbonization, sometimes called "wet" torrefaction.

Ribeiro et al. note that torrefaction is a more complex process than initially anticipated" and states that torrefaction of biomass is still an experimental technology Michael Wild, president of the International Biomass Torrefaction Council, stated in 2015 that the torrefaction sector is in its optimisation phase, i.e. it is maturing. He mentions process integration, energy and mass efficiency, mechanical compression and product quality as the variables most important to master at this point in the sector's development.

Environmental Impacts

GHG Savings

1. Yield and Soil Carbon Content:

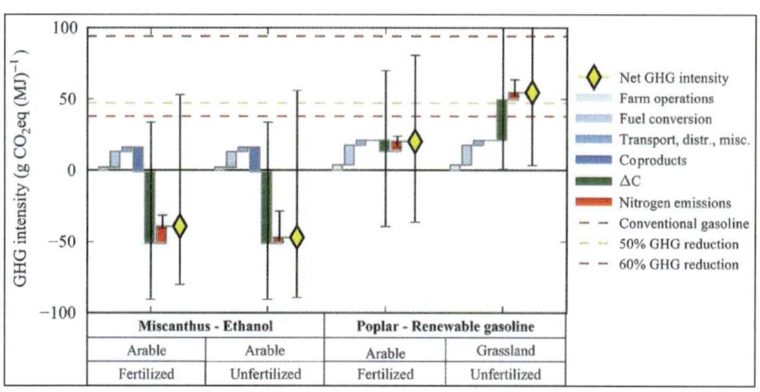

GHG/CO_2/carbon negativity for Miscanthus x giganteus production pathways.Relationship between above-ground yield (diagonal lines), soil organic carbon (X axis), and soil's potential for successful/unsuccessful carbon sequestration (Y axis). Basically, the higher the yield, the more land is usable as a GHG mitigation tool (including relatively carbon rich land.)

The amount of carbon sequestrated and the amount of GHG (greenhouse gases) emitted will determine if the total GHG life cycle cost of a bio-energy project is positive, neutral or negative. Specifically, a GHG/carbon negative life cycle is possible if the total below-ground carbon accumulation more than compensates for the above-ground total life-cycle GHG emissions. Whitaker et al. estimates that for Miscanthus x giganteus, carbon neutrality and even negativity is within reach. Basically, the yield and related carbon sequestration is so high that it more than compensates for both farm operations emissions, fuel conversion emissions and transport emissions. The graphic on the right displays two CO_2 negative Miscanthus x giganteus production pathways, represented in gram CO_2-equivalents per megajoule. The yellow diamonds represent mean values.

One should note that successful sequestration is dependent on planting sites, as the best soils for sequestration are those that are currently low in carbon. The varied results displayed in the graph highlights this fact. Milner et al. argues that for the UK, successful sequestration is expected for arable land over most of England and Wales, with unsuccessful sequestration expected in parts of Scotland, due to already carbon rich soils (existing woodland). Also, for Scotland, the relatively lower yields in this colder climate makes CO_2 negativity harder to achieve. Soils already rich in carbon includes peatland and mature forest. Grassland can also be carbon rich, however Milner et al. further argues that the most successful carbon sequestration in the UK takes place below improved grasslands. The bottom graphic displays the estimated yield necessary to achieve CO_2 negativity for different levels of existing soil carbon saturation.

The perennial rather than annual nature of Miscanthus crops implies that the significant below-ground carbon accumulation each year is allowed to continue undisturbed. No annual plowing or digging means no increased carbon oxidation and no stimulation of the microbe populations in the soil, and therefore no accelerated carbon-to-CO_2 conversion happening in the soil every spring.

2. Savings Comparison:

Fundamentally, the below-ground carbon accumulation works as a GHG mitigation tool because it removes CO_2 from the above-ground CO_2 circulation (the circulation from plant to atmosphere and back into plant.) The above-ground circulation is driven by photosynthesis and combustion – first, the miscanthus fields absorb CO_2 and assimilates it as carbon in its tissue both above and below ground. When the above-ground carbon is harvested and burned, it is released back into the atmosphere, in the form of CO_2. However, an equivalent amount of CO_2 (and possibly more, if the biomass is expanding) is absorbed back by next season's growth, and the cycle repeats. This above-ground cycle has the potential to be carbon neutral, but of course the human involvement in the process means additional energy input, often coming from fossil sources. If the fossil energy spent on the operation is high compared to the operation's energy output, the CO_2 footprint can approach, match or even exceed the CO_2 footprint from burning fossil fuels directly, as has been shown to be the case for several first-generation biofuel projects. Transport fuels might be worse than solid fuels in this regard.

The problem can be dealt with both from the perspective of increasing the amount of carbon that is moved below ground, and from the perspective of decreasing fossil fuel input to the above-ground operation. If enough carbon is moved below ground, it can compensate for a bio-energy project's total CO_2 emittance. On the other hand, if the above-ground CO_2 cost decreases, less below-ground carbon allocation is needed for the bioenergy project to become CO_2 neutral or negative.

For first generation bio-energy crops, the greenhouse gas footprints were often large, but second generation bio-energy crops like Miscanthus reduces its CO_2 footprint drastically. Hastings et al. found that Miscanthus crops almost always has a smaller environmental footprint than first generation annual bioenergy ones A large meta-study of 138 individual studies, done by Harris et al., revealed that second generation perennial grasses (miscanthus and switchgrass) on average stores 5 times more carbon in the ground than short rotation coppice forestry plantations (poplar and willow). Compared to fossil fuels, the GHG savings are large – without adding in the carbon sequestration negative numbers, Miscanthus fuel has a GHG cost of 0.4 - 1.6 grams CO_2-equivalents per megajoule, compared to 33 grams for coal, 22 for liquefied natural gas, 16 for North Sea gas, and 4 for wood chips imported to Britain from the USA.

Confirming the above numbers, McCalmont et al. found that the mean energy input/output ratios for Miscanthus is 10 times better than for annual crops, while GHG costs are 20-30 times better than for fossil fuels. For instance, Miscanthus chips for heating saved 22.3 tonnes of CO_2 emittance per hectare per year in the UK (9 tonnes per acre), while maize for heating and power saved 6.3 (2.5 per acre). Rapeseed for biodiesel saved only 3.2 (1.3 per acre). Lewandowski et al. found that each hectare (2.47 acres) of Central European arable land planted with Miscanthus can reduce the atmospheric CO_2 level with up to 30.6 tonnes per year, save 429 GJ of fossil energy used each year, with 78 euros earned per tonne reduced CO_2 (2387 euros earned per hectare per year) – given that the biomass is produced and used locally (within 500 km / 310 miles). For Miscanthus planted on marginal land limited by cold temperatures (Moscow), the reduction in atmospheric CO_2 is estimated to be 19.2 tonnes per hectare per year (7.7 tonnes per acre), with fossil energy savings of 273 GJ per hectare per year (110 GJ per acre). For marginal land limited by drought (Turkey), the atmospheric CO_2 level can potentially be reduced with 24 tonnes per hectare per year (9.7 tonnes per acre), with fossil energy savings of 338 GJ per hectare per year (137 tonnes per acre). Based on similar numbers, Poeplau and Don expect Miscanthus plantations to grow large in Europe in the coming decades. Whitaker et al. states that after some discussion, there is now consensus in the scientific community that the GHG balance of perennial bioenergy crop cultivation will often be favourable, also when considering the implicit direct and indirect land use changes.

Biodiversity

Felten and Emmerling found that the number of earthworm species per square meter was 5.1 for Miscanthus, 3 for maize, and 6,4 for fallow (totally unattended land), and states that it was clearly found that land-use intensity was the dominant regressor for earthworm abundance and total number of species." Because the extensive leaf litter on the ground helps the soil to stay moist, and also protect from predators, they conclude that Miscanthus had quite positive effects on earthworm communities and recommend that Miscanthus may facilitate a diverse earthworm community even in intensive agricultural landscapes."

Nsanganwimana et al. found that the bacterial activity of certain bacteria belonging to the proteobacteria group almost doubles in the presence of M. x giganteus root exudates.

Lewandowski et al. found that young Miscanthus stands sustain high plant species diversity, but as the Miscanthus stands mature, the canopy closes, and less sunlight reach the ground. In this situation it gets harder for the weeds to survive. Lewandowski et al. found 16 different weed species per 25 m2 plot. The dense canopy works as protection for other life-forms though; Lewandowski et al. notes that Miscanthus stands are usually reported to support farm biodiversity, providing habitat for birds, insects, and small mammals Both Haughton et al. and Bellamy et al. found that the Miscanthus overwinter vegetative structure provided an important cover and habitat resource, with high levels of diversity in comparison with competing energy grasses. This effect was particularly evident for beetles, flies, and birds, with breeding skylarks and lapwings being recorded in the crop itself. The Miscanthus crop offers a different ecological niche for each season – the authors attribute this to the continually evolving structural heterogeneity of a Miscanthus crop, with different species finding shelter at different times during its development – woodland birds found shelter in the winter and farmland birds in the summer. For birds, 0.92 breeding pairs species per hectare (0.37 per acre) was found in the Miscanthus field, compared to 0.28 (0.11) in the wheat field. The authors note that due to the high carbon to nitrogen ratio it is in the field's margins and interspersed woodlands that the majority of the food resoures are to be found. Miscanthus fields work as barriers against chemical leaching into these key habitats however.

Water Quality

McCalmont et al. claims Miscanthus fields leads to significantly improved water quality because of significantly less nitrate leaching. Likewise, Whitaker et al. claims that there is drastically reduced nitrate leaching from Miscanthus fields compared to the typical maize/soy rotation because of low or zero fertilizer requirements, the continuous presence of a plant root sink for nitrogen, and the efficient internal recycling of nutrients by perennial grass species. For instance, a recent meta-study concluded that Miscanthus had nine times less subsurface loss of nitrate compared to maize or maize grown in rotation with soya bean.

Soil Quality

The fibrous, extensive Miscanthus rooting system and the lack of tillage disturbance improves infiltration, hydraulic conductivity and water storage compared to annual row crops, and results in the porous and low bulk density soil typical under perennial grasses, with water holding capabilities expected to increase by 100-150 mm. Nsanganwimana et al. argues that Miscanthus improves carbon input to the soil, and promote microorganism activity and diversity, which are important for soil particle aggregation and rehabilitation processes. On a former fly ash deposit site, with alkaline pH, nutrient deficiency, and little water-holding capacity, a Miscanthus x giganteus crop was successfully established – in the sense that the roots and rhizomes grew quite well, supporting and enhancing nitrification processes, although the above-ground dry weight yield was low because of the conditions. The authors argue that Miscanthus' ability to improve soil quality, even on contaminated land, can be used in combination with organic amendments on soils with a low agronomic value. For instance, there is a great potential to increase yield on contaminated marginal land low in nutrients by fertilizing it with nutrient-rich sewage sludge or wastewater. The authors claim that this practice offer the three-fold advantage of improving soil productivity, increasing biomass yields, and reducing costs for treatment and disposal of sewage sludge in line with the specific legislation in each country.

Invasiveness

Miscanthus x giganteus' parents on both sides, M. sinensis and M. sacchariflorus, are both potentially invasive species, because they both produce viable seeds. M. x giganteus does not produce viable seeds however, and Nsanganwimana et al. claims that there has been no report on the threat of invasion due to rhizome growth extension from long-term commercial plantations to neighboring arable land."

LANDFILL GAS

Landfill gas is a complex mix of different gases created by the action of microorganisms within a landfill. Landfill gas is approximately forty to sixty percent methane, with the remainder being mostly carbon dioxide. Trace amounts of other volatile organic compounds (VOCs) comprise the remainder (<1%). These trace gases include a large array of species, mainly simple hydrocarbons.

Landfill gases have an influence on climate change. The major components are CO_2 and methane, both of which are greenhouse gas. Methane in the atmosphere is a far more potent greenhouse gas, with each molecule having twenty five times the effect of a molecule of carbon dioxide. Methane itself however accounts for less composition of the atmosphere than does carbon dioxide. Landfills are the third largest source of methane in the US.

Production

Landfill gases are the result of three processes:

- Evaporation of volatile organic compounds (e.g., solvents).
- Chemical reactions between waste components.
- Microbial action, especially methanogenesis.

The first two depend strongly on the nature of the waste. The dominant process in most landfills is the third process whereby anaerobic bacteria decompose organic waste to produce biogas, which consists of methane and carbon dioxide together with traces of other compounds. Despite the heterogeneity of waste, the evolution of gases follows well defined kinetic pattern. Formation of methane and CO_2 commences about six months after depositing the landfill material. The evolution of gas reaches a maximum at about 20 years, then declines over the course of decades.

Monitoring

Because gases produced by landfills are both valuable and sometimes hazardous, monitoring techniques have been developed. Flame ionization detectors can be used to measure methane levels as well as total VOC levels. Surface monitoring and sub-surface monitoring as well as monitoring of the ambient air is carried out. In the U.S., under the Clean Air Act of 1990, it is required that many large landfills install gas collection and control systems, which means that at the very least the facilities must collect and flare the gas.

U.S. Federal regulations under Subtitle D of RCRA formed in October 1979 regulate the siting, design, construction, operation, monitoring, and closure of MSW landfills. Subtitle D now requires controls on the migration of methane in landfill gas. Monitoring requirements must be met at landfills during their operation, and for an additional 30 years after. The landfills affected by Subtitle D of RCRA are required to control gas by establishing a way to check for methane emissions periodically and therefore prevent off-site migration. Landfill owners and operators must make sure the concentration of methane gas does not exceed 25% of the LEL for methane in the facilities' structures and the LEL for methane at the facility boundary.

Use

A United States Environmental Protection Agency (EPA) report indicates that as of 2016, counts of operational municipal solid waste landfills range between 1,900 and 2,000. In a nation-wide study done by the Environmental Research and Education Foundation in 2013, only 1,540 operational municipal solid waste landfills were counted throughout the United States. Decomposing waste in these landfills produce landfill gas, which is a mixture of about half methane and half carbon dioxide. Landfills are the third largest source of methane emissions in the United States, with municipal solid waste landfills representing 95 percent of this fraction.

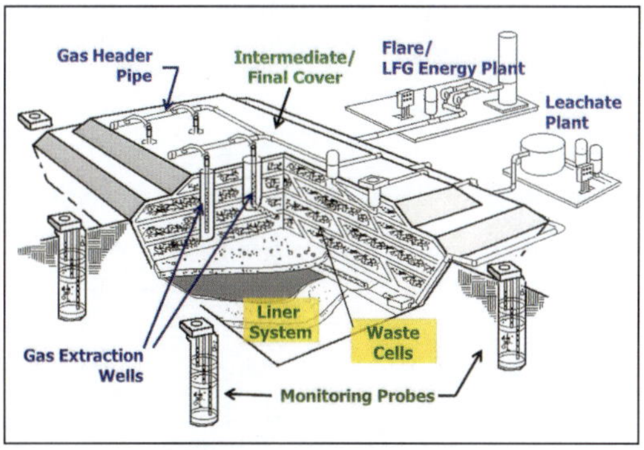

Landfill gas collection system.

The gases produced within a landfill can be collected and used in various ways. The landfill gas can be utilized directly on site by a boiler or any type of combustion system, providing heat. Electricity can also be generated on site through the use of microturbines, steam turbines, or fuel cells. The landfill gas can also be sold off site and sent into natural gas pipelines. This approach requires the gas to be processed into pipeline quality, e.g., by removing various contaminants and components. The efficiency of gas collection at landfills directly impacts the amount of energy that can be recovered - closed landfills (those no longer accepting waste) collect gas more efficiently than open landfills (those that are still accepting waste). A comparison of collection efficiency at closed and open landfills found about a 17 percentage point difference between the two.

Landfill gas can also be used to evaporate leachate, another byproduct of the landfill process. This application displaces another fuel that was previously used for the same thing.

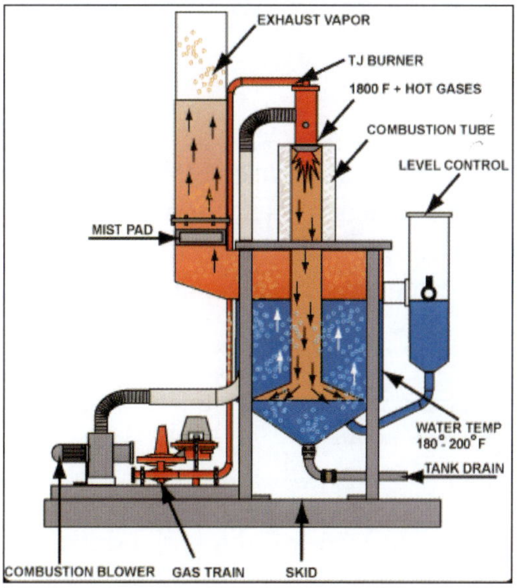

Leachate evaporation system.

In the U.S., the number of landfill gas projects increased from 399 in 2005, to 594 in 2012 according to the Environmental Protection Agency. These projects are popular because they control energy costs and reduce greenhouse gas emissions. These projects collect the methane gas and treat it, so it can be used for electricity or upgraded to pipeline-grade gas. (Methane gas has twenty-one times the global warming potential of carbon dioxide). For example, in the U.S., Waste Management uses landfill gas as an energy source at 110 landfill gas-to-energy facilities. This energy production offsets almost two million tons of coal per year, creating energy equivalent to that needed by four hundred thousand homes. These projects also reduce greenhouse gas emissions into the atmosphere.

The EPA, which estimates that hundreds of landfills could support gas to energy projects, has also established the Landfill Methane Outreach Program. This program was developed to reduce methane emissions from landfills in a cost-effective manner by encouraging the development of environmentally and economically beneficial landfill gas-to-energy projects.

Opposition

Capture and use of landfill gas can be expensive. Some environmental groups claim that the projects do not produce "renewable power" because trash (their source) is not renewable. The Sierra Club opposes government subsidies for such projects. The Natural Resources Defense Council (NRDC) argues that government incentives should be directed more towards solar, wind, and energy-efficiency efforts.

Microbial Oxidation

When landfill gas permeates through a soil cover, a fraction of the methane in the gas is oxidized microbially to CO_2.

Safety

Landfill gas emissions can lead to environmental, hygiene and security problems in the landfill. Several accidents have occurred, for example at Loscoe, England in 1986, where migrating landfill gas accumulated and partially destroyed a property. An accident causing two deaths occurred from an explosion in a house adjacent to Skellingsted landfill in Denmark in 1991. Due to the risk presented by landfill gas, there is a clear need to monitor gas produced by landfills. In addition to the risk of fire and explosion, gas migration in the subsurface can result in contact of landfill gas with groundwater. This, in turn, can result in contamination of groundwater by organic compounds present in nearly all landfill gas.

Although usually evolved only in trace amounts, landfills do release some aromatics and chlorocarbons.

Landfill gas migration, due to pressure differentials and diffusion, can occur. This can create an explosion hazard if the gas reaches sufficiently high concentrations in adjacent buildings.

STARCH CROPS

Starch-based feedstocks include grains, such as corn or wheat, and tubers such as (sweet) potatoes and cassava. These feedstocks contain long complex chains of sugar molecules. The starch can easily be converted to fermentable sugars. The sugar can then be converted to ethanol or drop-in fuels. The fibrous part of the plants (e.g. wheat straw or corn stover) can be converted to advanced biofuels.

In Europe, wheat is currently the main starch crop for bioethanol production. 0.7% of EU agricultural land and 2% of Europe's grain supply is used for production of renewable ethanol. The EC has proposed to limit biofuel produced from "food crops" at 7% of energy use in transport, due to concerns about food price and land use impacts. However, recent reports (and the latest statistics) suggest that the impacts of biofuels production from starch crops may have been greatly exaggerated and the many benefits of biofuels (European fuel security, job and wealth creation, production of valuable byproducts, GHG reduction) have not been fully taken into account.

Starch Crop Species

Wheat (Triticum Species)

After corn and rice, wheat is the most produced crop worldwide with a starch content of about 70%. It is a major food crop but can also be converted to bioethanol. 2.8 million metric tons of wheat were processed to ethanol in 2014. In the UK, Vivergo opened a 420 Mio litre wheat-to-ethanol plant in 2013 that also produces high protein animal feed as a by product.

Corn (Zea Mays)

Corn contains about 70% starch. In 2014 5.2 million metric tons of corn served the bioethanol business in Europe. Corn is the feedstock for more than 90% of ethanol production in the United States due to its abundance and low price; most ethanol is produced in the corn-growing states of the Midwest. The economic output of the 'renewable fuels industry' is $184 billion. It supports over 852,000 jobs and $56 billion in wages and generates about $14.5 billion in local and state tax revenue every year.

Barley (Hordeum Vulgare)

Barley is a winter crop that is planted in rotation with crops such as corn and soybean. It therefore helps to anchor the soil during a time when fields would otherwise lie fallow. It requires much less nitrogen fertilizer than, for example, corn. It shows potential as a biofuel feedstock, particularly in regions where the market for barley is not so big. 541,000 metric tons of barley were used as feedstock for ethanol in Europe in 2014. The starch content of barley varies between 50 and 75 %.

Rye (Secale Sereale)

Rye is a rather robust grain that also grows on poorer soils. Its starch content is about 60 %. In 2014 846,000 metric tons served as biofuel feedstock in Europe.

Millet/Sorghum (Species of e.g. Sorghum, Penisetum and Panicum)

Millet/Sorghum species can grow on marginal soils, need low inputs and improve soil productivity, and can be used in crop rotation systems as an ameliorant. Their starch content is about 75 %. In 2012, a study by the University of Nebraska–Lincoln compared ethanol production from millet cultivars and advanced breeding lines with that from normal corn and "highly fermentable" corn. Fermentation efficiencies in the tested millet varieties ranged from 84 percent to 91 per cent compared to 97 per cent from the highly fermentable corn hybrid.

Potatoes (Solanum Tuberosum)

Potatoes contain up to 19 % starch. In recent years, the food versus fuel debate, has focused attention on the use of waste from potato processing industries as a biofuels feedstock.

Cassava (Manihot Esculenta)

Cassava is an important food and feed crop in many tropical countries. It can also be cultivated on drier or poorer soils. Under good gowing conditions, yields can be as high as 150 tonnes/ha. Cassava contains circa 40 % of starch. China is a big promoter of cassava as biofuel feedstock. The USDA gain report published that in Thailand, the amount of ethanol produced from cassava was expected to double to 3 million tonnes in 2015. In October 2013, China New Energy Ltd announced a partnership with Sunbird Bioenergy Africa Ltd to build a cassva-to-bioenergy refinery in Nigeria (with a potential 9 more to follow). Mozambique also has a cassava-to-ethanol facility in operation.

Sweet Potatoes (Ipomoea Batatas)

Sweet potatoes can be cultivated in tropical or warm regions and offer relatively high ethanol yields since their starch content is about 70 %. The potential of sweet potatoes as a feedstock has been investigated for many decades. A study in 2008 suggested sweet potatoes may offer three times the yield of corn (in terms of ethanol per hectare). However, researchers in other studies have suggested that production costs of ethanol from sweet potato are "ten times those for corn". Worldwide, China is the biggest producer of sweet potatoes, and is active in their conversion to fuel ethanol

MUNICIPAL SOLID WASTE

The solutions of MSW management should not only be environmentally sustainable but also cost-efficient and socially acceptable. There are several factors that influence this complex process, which are largely intertwined. First of all, there is a need for political will, a willingness to pursue changes. While the EU sets a direction (i.e. a shift towards the circular economy), waste management is implemented at national level. Waste management contains a multi-level governance system, embracing central governments for setting strategies, creating national plans, then regional and in most cases local authorities for designing and implementing policies and organising tools for waste collection, treatment, and disposal. Government regulations, taxes and support schemes play an important part in the development of this sector and encouragement of new technologies, which would struggle to survive without governmental support. Even though 'Advances in Technology' are essential tools to facilitate change, there should be a platform for innovation and technology demand. Sometimes regulatory barriers may prevent the commercialisation of new beneficial technologies. Equally, there has to be a vivid business case for the industry to get involved.

Private investment played an essential role in the development of the Estonian waste management system. Markets for 'secondary' waste-derived raw materials need to be created. From an economic point of view, new technologies typically face an uphill struggle when first introduced due to a lack of economies of scale and learning economies, not yet established supply chains and market structures. Depending on the technologies involved, they may not fit neatly with the existing infrastructure. While the WtE sector presents an immense business potential with financial value in new circular business models: enabling companies to make substantial economic gains and become more competitive, the obstacles defined above must be overcome. The challenges of municipal waste management also stem from the direct proximity of the generated waste to citizens, a very high public visibility, and an active involvement of citizens and businesses (i.e. willingness to recycle etc.). Different countries have different societal structures that interject in part to different 'societal preferences'. For instance, a WtE installation may face effective local protest (known as a NIMBY (Not In My Backyard) syndrome) especially if a sufficient case for nuisance through noise or odours can be made. Ren et al. in their study analysed risk perception and public acceptance over protested WtE facilities. According to their findings, WtE processes should be promoted through highlighting their benefits as renewable energy sources and the conservation of land to weaken strong protests from local communities against incineration plants, especially in areas with high population densities. The most significant factors influencing the choice of protest were: risk perception, income, opinion about the benefits of WtE, gender and previous experience of odour pollution. Additionally, a significant statistical relationship between knowledge and risk perception was identified. Thus, the need for risk communication, as well as involving the public in the whole management process, is highly recommended. New technologies also quite often require significant changes in user practices, habits, and aspirations. A new term of 'prosumer' has emerged attributing to the role of active consumers with the potential to be energy producers, particularly through self-generation of renewable energy, storage, energy conservation and participation in demand response. Waste management and its impact on the environment and human health is self-evident: poor choices, such as landfills may, depending on the way they are built, contaminate soil and water with chemicals contained in waste and also lead to a climate change.

Table: Factors that influence waste management.

Factors of Waste Management	
Political	Political will, multi-level governance, government regulations (taxes, subsidies), data collection and monitoring.
Economic	Business model, cost-benefit analysis, availability of finance, collaboration, and transparency along the value chain.
Environmental	Sustainability policy, human health impact.
Social	Community perception.
Technological advances	Innovation, infrastructure.
Educational	Research centres, cooperation projects.

Finally, education, public awareness and advocacy on the organisation of waste management and the potential of MSW to be used as an energy resource should be not forgotten. The cooperation of the institutions (research centres and Ministries, such as Environment and Energy) within and among Member States should be encouraged. For instance, a regional project "RECO Baltic 21-Tech" serves as an inspiring example of mutual benefits: it encompassed universities, research centres and companies, which work hand in hand in search of solutions for the problems associated with waste management in Baltic Sea region countries, such as Germany, Poland, Lithuania, Estonia, Latvia and Sweden, contributing to sustainable waste management in the region, at the same time gaining interesting topics for research, possibilities for capacity building and curriculum enrichment for students. Along similar lines, there is another three-year European project COOLSWEEP (funded by the EU 7th Framework programme), involving six different European regions. The COOLSWEEP project seeks to identify possible synergies and partnerships between public and private stakeholders and knowledge institutions within the field of WtE and to support the development of new businesses based on utilising waste as the feed material for efficient sustainable energy production and more effective use of the by-products from the energy generation. COOLSWEEP also has an additional mentoring task to establish a new cluster based in Riga, Latvia, with strong competencies within the field of WtE. This is an essential element for the WtE development in Latvia given its current underdeveloped WtE infrastructure.

The preparation and organisation of MSW management with its recent trend centred on the circular economy needs inputs from a range of disciplines, therefore, we embrace an inter-disciplinary perspective (legal, scientific, and to a lesser extent economic). Previous studies on waste management either had a scientific focus, or economic justifications, socio-legal aspects or technology advancements and related issues, including WtE plants in some European countries. There is also a profound literature on the future of renewable energy and sustainability compiled by Professor Olabi with an emphasis being placed on energy security.

There have been some studies with a specific emphasis on WtE, for instance, under the influence of market and EU legislation. The most recent broadest (in scope) study was conducted by the European Environment Agency "Assessment of waste incineration capacity and waste shipments in Europe" in 2017, with the previous study on incineration overcapacity and waste shipping in Europe commissioned by the Global Alliance for Incinerator Alternatives in 2013. These studies observed an uneven distribution of WtE capacity across Europe with the six countries – Germany, France, the Netherlands, Italy, the United Kingdom, and Sweden - accounting for almost three-quarters of Europe's incineration capacity and with the other countries heavily relying on landfill for MSW disposal. These reports also noted a lack of consistent data across Europe.

Further studies focused on different aspects of waste management in specific countries. For instance, Guziana et al. focused on the EU waste management and energy policy in the context of Sweden. They pointed out, that nowadays the priority of waste prevention and concern for food losses has significantly influenced the WtE sector in the EU: the amount of waste delivered to incineration plants will gradually decrease since the recycling rate and waste prevention will grow. They concluded that there is a need to redesign waste management systems in order to meet the waste hierarchy.

Given the limited interdisciplinary studies on waste management (and WtE as an important part of it) as well as a narrow scope of jurisdictional coverage, we aim to fill this gap in the literature. Building on the previous studies, we selected ten countries (i.e. Estonia, Greece, Italy, Latvia, Lithuania, Norway, Poland, Slovenia, Spain, and UK) to better represent a European context. These ten countries were chosen because they offer a range of perspectives in terms of their different stages of economic development (e.g. the UK and Norway with a GDP per capita above the EU average (EA-106), and the rest of the countries being below this level, as shown in Table); different sizes of their economies (small, for example, Estonia, Latvia, Lithuania and Slovenia and large - the UK); different Accession waves (EU and EEA (European Economic Area) countries): Italy as one of the co-founding states of the EU; Estonia, Latvia, Lithuania, Poland, and Slovenia (i.e. the newer Member States which joined the EU in 2004), the UK – a potential leaver of the EU, and Norway as a representation of an EEA country; and, finally, geographical positioning: Northern (Norway), Southern (Greece, Italy, Spain), Eastern and Central European states (Estonia, Latvia, Lithuania, Poland, and Slovenia) and Western (UK). The last groupings will be followed here, as they represent not only geographical regions, but also similarities in the countries' GDP (with the exception of Greece due to the State's debts); and duration of the EU membership – the Southern region countries classified as 'old' Member States, the Central European region – falling to the 'newer' Member States classification, the Northern region – representing the EEA rather than being a Member State of the EU and finally, the Western region – referring to the Member State, whose fate is not clear due to the Brexit negotiations.

Table: Profile of the selected countries.

GEO	Population (mil)	Size/Km²	Density	GDP per capita	Administrative units
Estonia	1.315	45,227	28.8	74	213 municipalities: 183 rural municipalities and 30 cities.
Greece	10.784	131,957	86.4	67	13 regions.
Italy	62.808	302,073	201.32	96	8040 municipalities.
Latvia	1.95	64,573	30,19	65	10 waste management regions/110 municipalities and 9 independent cities.
Lithuania	2.88	65,286	44.1	75	10 counties/counties are subdivided into 60 municipalities.
Norway	5.267	304,282	14	149	19 administrative regions/426 municipalities.
Poland	38.42	312,679	123	69	2479 municipalities.
Slovenia	2.06	20,273	102	83	212 municipalities/8 regional MSW management centres.
Spain	46.446	505,944	92	92	17 autonomous communities, 2 cities with statutes of autonomy, and 8125 local institutions.
UK	65.6	248,528	250	108	England: 353 local authorities; Wales: 22 and Scotland: 32 county councils; Northern Ireland: 11 district councils .

The duration of EU membership does not seem to correlate with the states' waste management performance, as it depends on different visions, strategies, and priorities of waste management. Yet, historical data of some Member States (for instance, the Baltic states) are not available due to different classifications of waste streams before joining the EU. Therefore, the data here aim to cover the 2004–2015 period (and 2016 where possible) with the template provided for each national rapporteur. While national rapporteurs collected the data mainly drawn from the NWMPs and other governmental reports, statistical data from Eurostat and the EU reports prepared for the EEA (European Environment Agency), and for the European Commission (BiPRO) were also consulted. Information and the availability of data in this field is rather limited. It has been noted that the Member States have difficulties in collecting and calculating the necessary information on MSW for the EU reporting. For instance, the three Baltic countries have producer responsibility schemes for packaging waste. Given that there is no established reporting system, private operators of these schemes do not always provide data on the sources of the packaging waste, and therefore packaging waste is not always included in Eurostat as MSW in the Baltic countries. Similar situation exists in other countries, such as Greece. There is also no uniformity with regard to MSW sent to Mechanical Biological Treatment (MBT); in some countries the whole amount received at the MBT plant is allocated to recycling, whereas in other countries – only the actual amount recycled after the MBT.

MSW Management

Waste management is a complex field which goes beyond prevention, collection, treatment, and disposal of waste, embracing a larger ambit of socioeconomic development, government regulations, policy choices, and resource management. It is about protecting, preserving, and improving the quality of the environment, human health, ensuring prudent and rational utilisation of natural resources, promoting a more circular economy, through improving resource use and the efficiency of such use and by ensuring waste is valued as a resource. The European Commission expressed that the EU's economy currently loses a significant amount of potential secondary raw materials which are found in waste streams, with only a limited share (43%) of the municipal waste generated in the Union being recycled, with the rest being landfilled (31%) or incinerated (26%). Given that the EU misses out significant opportunities to reclaim as many resources as possible and to improve resource efficiency and to enhance the transition towards a circular economy, it sets the targets for the Member States and for the Union itself.

The European Parliament, which urged the Commission to put forward ambitious proposals to revise waste legislation, plays an essential role in the EU's transition towards a circular economy. In March 2017, the European Parliament adopted its position on the review of the four key waste directives with more ambitious targets and provisions compared to the Commission's proposals issued in 2016. For example, in some of its 234 proposed amendments to the WFD and to accelerate the switch to a circular economy, the Parliament tightened the Commission's proposed targets for the preparation for reuse and recycling of municipal solid waste to at least 60% by 2025 (including a minimum of 3% of total municipal waste prepared for re-use) and at least 70% by 2030 (including a minimum of 5% of total municipal waste prepared for re-use), and also called on the Commission to examine the opportunity to set EU waste prevention targets. Furthermore, these amendments set stricter obligations for separate collection, removing Commission text that this should be carried out where 'technically, environmentally and economically practicable'. The Parliament also urged the Commission to introduce a food waste hierarchy as well as to set binding food waste reduction targets.

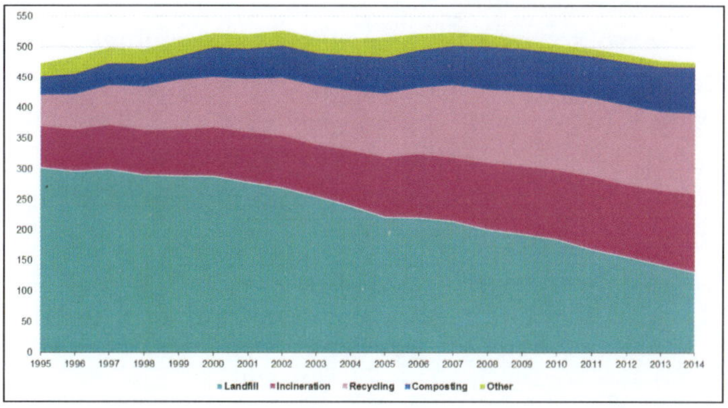

Municipal waste treatment in EU-27 by type of treatment, (kg per capita), 1995–2015.

The Parliament's amendments to the Landfill directive also call for a maximum of 5% of total MSW to be sent to landfills in 2030, compared to 10% in the Commission's proposal. The Member States are free to set even tighter targets.

Waste management planning is one of the key tools for authorities to convert the principles of EU waste legislation at national, regional, and local level within their Member State. The Member States' authorities are obliged to establish one or more waste management plans (WMP) pursuant to Articles 1, 4, 13 and 16 of the WFD. These must cover the entire geographical territory of the Member State concerned, set out an analysis of the current waste management situation, as well as the measures to be taken with respect to environmentally sound preparation for re-use, recycling, recovery and disposal of waste, and an evaluation of how the plan will support the implementation of the objectives and provisions of the Directive. The waste hierarchy is a good guideline to assess waste management options, particularly when waste management plans are being developed or reviewed at the national or regional level. Any revisions of the WMP must be notified to the Commission, as they play a key role in identifying any gaps in the fulfilment of the requirements and objectives of the EU waste legislation. The European Commission can pursue legal action against the Member States for their failure to comply with waste management obligations under Article 258 TFEU. For instance, the Commission has issued final warnings to Romania for failing to review and adopt its national waste management plan and waste prevention programme, in line with the objectives of EU WFD and the circular economy. The Commission is also forewarning to take Slovenia to the Court of Justice for its alleged failure to close and rehabilitate 28 illegal landfills which represent a serious risk for human health and the environment. The European Commission has also brought several infringement cases against Estonia concerning EU waste directives, including referral to the Court of Justice for incomplete or unreported transposition of the three directives: the Landfill Directive (Case C-195/10), the Mining Waste Directive (Case C-515/10) and the Directive on Waste Electrical and Electronic Equipment (Case C-528/09). These cases are closed now, except for a formal notice concerning the Directive on lightweight plastic carrier bags.

Waste-to-Energy: Role of WtE in the Circular Economy

One may argue that waste to energy can encourage wastefulness and discourage recycling to ensure regular feedstock to incinerators, which do not have a good reputation due to released toxins and greenhouse gases. Along similar lines, some NGOs, such as Zero Waste Europe (ZWE), declare that conceptually speaking WtE does not have a place in the circular economy as the material loops are closed when 'there is nothing left to burn'. Yet, this conclusion is made solely in the context of waste to energy incineration. The European Commission in its recently published communication, which is meant to clarify the role of WtE in the circular economy, expressed that:

> "Waste-to-energy processes can play a role in the transition to a circular economy provided that the EU waste hierarchy is used as a guiding principle

and that choices made do not prevent higher levels of prevention, reuse and recycling."

However, there is a rather sceptical view towards WtE, as the Commission is concerned that, by increasing WtE capacity, recycling will be jeopardised, hence undermining the waste hierarchy. Even though the Commission concludes that WtE could play a role in the transition to a circular economy, the waste hierarchy must be used as a guiding principle to ensure that prevention, reuse, and recycling are not averted. Most certainly, WtE plants meet the requirements set by the Industrial Emissions Directive. In addition, the Commission notes that the Member States in their future waste management plans to invest in WtE must take into account the risk of "stranded assets" and new plants should only be built provided the availability of feedstock would be sustained for the operation of new incineration plants over their lifespan (20–30 years) without neglecting separate collection and recycling obligations. Finally, the Member States are also advised to gradually phase-out public support for the recovery of energy from mixed waste and either introduce or raise incineration taxes. Therefore, the message from the EU is clear that the development of separate collection structures and recycling capacity (preferably in the form of anaerobic digestion) should be a priority.

While there is some scepticism towards WtE from a waste management point of view, there is a different story from a renewable energy point of view. biomass (which embraces the biodegradable fraction of municipal waste) is one of the renewable energy sources defined by the RED. Biomass is encouraged as a renewable energy source in the newest proposal of REDII. 'Waste' can cease to be a problem and become a valuable resource. The inclusion of the organic portion of MSW in the definition of potential sources of renewable energy has enabled the Member States to meet their national renewable energy targets via the WtE incineration industry. Statistically speaking, biomass and waste are the largest sources of 'renewable energy' in Europe amounting to 63.1% of the total share of renewable energy sources. The Commission in November 2016 published the "Clean Energy for all Europeans" strategy called "Winter package", where among eight legislative instruments, the Commission recommends setting a new target of at least 27% renewables by 2030 pursuant to REDII. Yet, the Parliament has reiterated its request for a binding target of at least 30% of total energy consumption coming from renewable energy sources by 2030, combined with binding national targets. Therefore, biomass as a renewable source would enable the Member States to meet the constantly increasing EU energy target from renewables.

Waste Hierarchy and Technology

Even though not legally binding, the recent communication on the role of waste-to-energy in the circular economy highlighted that the recovery of energy from waste supports the EU objectives of the circular economy action plan and is firmly guided by the waste hierarchy. Some WtE processes are clearly assigned to various steps in the waste hierarchy, which is shown in Fig. It is expected that these processes will be transposed into the revised WFD to have more credibility.

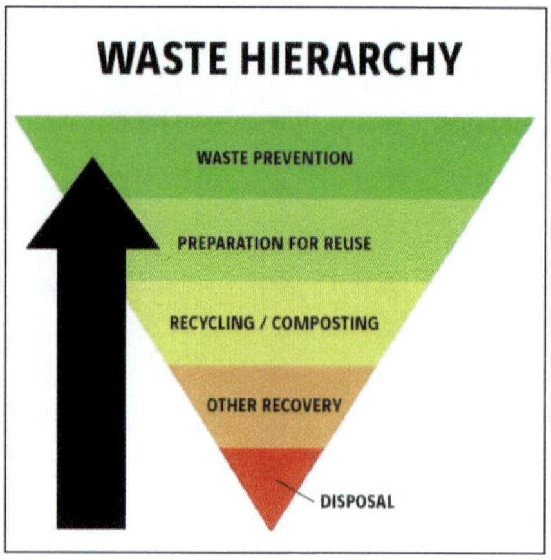

Waste-to-Energy in the waste hierarchy.

As can be seen, incineration with high efficient energy recovery is understood as waste recovery, but anaerobic digestion is considered as recycling. Regrettably, the communication sends a message that WtE should be avoided as much as possible leaving it to the stage where recycling is no longer possible. Additionally, it provides some guidance to the Member States on how to improve use of economic instruments and capacity planning to avoid potential overcapacity of incineration infrastructure. Finally, the best available technologies and processes with high energy efficiency are also listed. The European Commission recommends several technologies.

Co-incineration in combustion plants: gasification of solid recovered fuel (SRF) and co-incineration of the resulting syngas in the combustion plant:

- Co-incineration in cement kilns;
- Incineration in dedicated facilities:
    - The use of super heaters and heat pumps;
    - The utilisation of the energy contained in flue gas;
    - Distributing chilled water through district cooling networks; and heat through low temperature district heat networks.
- Anaerobic digestion with upgrading of the biogas into bio-methane for further distribution and utilisation.

Pursuant to the waste hierarchy defined in the WFD, WtE can be attributed to either a 'disposal' classification or 'other recovery' depending of the level of energy recovered. The energy efficiency (EE) of the installation must be ≥ 0.65 for facilities in operation since 2009 and ≥ 0.60 for facilities in operation before 2009. The EE is calculated following the equation,

$$EE = \frac{[Ep - E + Ei]}{[0.97 \times (Ew + Ef)]}$$

where,

- EE = Energy efficiency.
- Ep = Energy produced (electricity or heat) in GJ/year.
- Ef = Energy consumption as fuel in GJ/year.
- Ew = Energy content of wastes in GJ/year.
- Ei = Annual imported energy excluding Ew and Ef in GJ/year.

All Member States must follow this formula for their WtE plants classification while aiming to achieve a higher rank in the waste hierarchy.

Existing Waste-to-Energy Technologies

A number of new market technologies, such as anaerobic digestion, pyrolysis and gasification, are in the process of being deployed. These technologies provide the potential to recover products from the waste stream which complete incineration would not allow and a significant proportion focuses on biomass waste.

Waste-to-energy technologies may be divided into three main groups based on the conversion process used: thermochemical, physicochemical, and biochemical. Modern technologies of thermochemical conversion include high-efficiency combustion, pyrolysis or gasification. Moreover, sanitary landfilling together with capture and utilisation of produced biogas is an important part of waste management, especially in developing countries, Bioethanol produced from waste via fermentation also can be used to produce energy. The compilation of available WtE technologies is shown in figure.

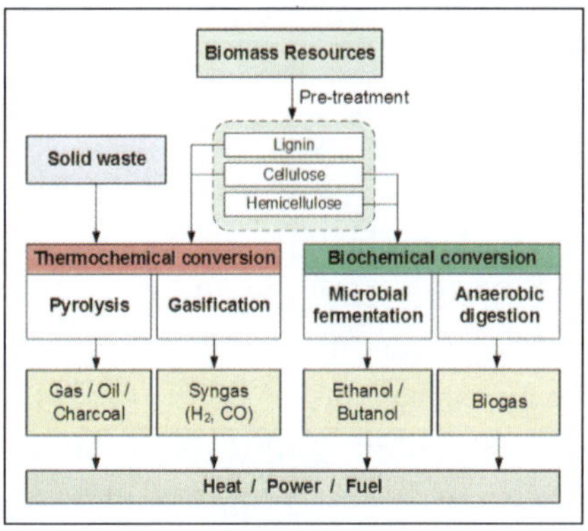

Waste-to-Energy technologies based on applied conversion process.

The heat generated by burning the waste should at a minimum be high enough to warrant proper combustion conditions and produce sufficient amounts of energy to overcome losses and auxiliary consumption. This is possible with a lower heating value of at least 4 MJ/kg. Modern waste-to-energy plants can export energy (usually heat and electricity) with very low environmental impact. A typical plant consists of four basic elements: combustor, recovery boiler, flue gas treatment system, and steam cycle. The hot gas generated in the combustor goes through the recovery boiler to produce steam, which can be used directly as a heat carrier or may be sent to a steam turbine to produce electricity. Finally, flue gases are treated by sorption and filtration.

The incineration of waste releases the energy fixed in them. However, to enable the transfer of energy to the circulating working medium, devices called boilers need to be used. Nowadays mainly two technologies are used in order to combust municipal waste: moving grate boilers and fluidized bed boilers. The moving grate boilers constitute the vast majority – they are used in 87% of plants in Europe. The technology is well known and investigated for a long time. For instance, this technology was chosen for all new-built WtE plants in Poland. The idea of such a boiler is shown in fig. A moving grate principle of operation consists in a slow movement of the processed waste in the layer, which is burned in subsequent phases as follows: drying, pyrolysis/gasification, combustion and burnout. The temperature of the gas resulting from the process has to be raised, after the last injection of combustion air, to 850 °C for two seconds (or to 1100 °C if hazardous wastes with a content of more than 1% of halogenated organic substances, expressed as chlorine, are incinerated).

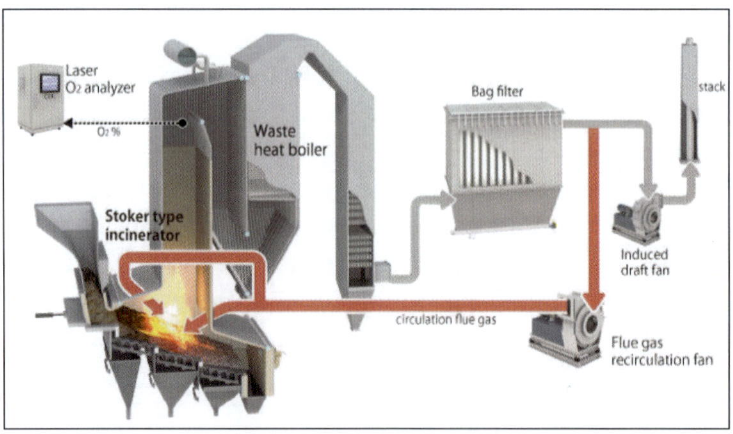

Scheme of moving grate boiler for waste incineration.

In a circular development model, materials and their value in circulation within the economic system are retained for as long as possible, by optimising the integrated waste cycle in order to put resources to efficient use. In the EU vocabulary re-use, recycling, and recovery are the key words that a new paradigm is built upon to promote sustainability, innovation, and competitiveness. multifunctional technology solutions presented by WtE should be given more attention in the circular economy due to their capacity to bridge and enhance resource and energy efficiency improvements. WtE has a role to play in the circular economy, contributing to synergies in three EU policies – waste management, energy union, and environmental (climate change) policy enabling the Member States to meet their targets linked to these policies, especially in the context of resources and energy efficiency. Given that the EU is the main driver of change in the EU Member States (including the EEA), recent changes to the legislative framework on waste management were also reviewed. It seems that the European Commission is set to provide further clarity on the notion of MSW to avoid any ambiguities among the Member States. Yet, further harmonisation at EU level is required, especially in the context of setting well-defined criteria for the 'end-of-waste' status and the calculation rules (i.e. there is currently no single homogenous method on how to calculate what is recycled, composted, or landfilled, which has caused some uncertainties, particularly in newer Member States). Among new legal initiatives, the first Commission document to define the role of WtE in the circular economy is welcome, despite not being legally binding. In this document, the Commission assigned some WtE processes to various steps in the waste hierarchy, where it also noted that WtE could play a role in the transition to a circular economy provided the waste hierarchy is used as a guiding principle to ensure that prevention, reuse, and recycling are not averted. Yet, the communication could have gone a step further assigning other technologies to the waste hierarchy. Indeed, apart from anaerobic digestion, there are a number of new market technologies, such as pyrolysis and gasification that provide the potential to recover products from the waste stream which complete incineration would not allow. These should be further utilised in the European countries.

The examination of waste management development in the selected European countries revealed that the biggest proportion of waste (above the EU average) was generated in Italy (486kg/per capita), Greece, and the UK (each 485kg/per capita). Even though GDP is higher in the UK and Italy, the same cannot be said about Greece (i.e. which together with Latvia has one of the lowest GDP rates per capita). The lack of correlation between GDP and waste generation can be noted in Norway; despite having the highest GDP rate, its waste generation was below EU average and some other countries analysed, such as Slovenia, Spain, and Lithuania. Poland generated a remarkably low amount of MSW in comparison to the other countries analysed. This can be surmised to be due to illegal dumping and, therefore, a failure to report waste generated. Additionally, the definition of MSW differs across the analysed countries in Europe. Some of them identify rather detailed potential sources of waste similar to household waste, whereas others (i.e. Estonia, Latvia) provide limited sources, referring to "similar to household waste from trade". Therefore, this can influence statistical data on the generation of MSW.

Despite attempts to transfer to waste-to-energy, landfilling is still a problem in most Member States, especially in Greece (81%) and Latvia (79%), followed by Lithuania and Spain (reaching 55% each). There is no surprise that two countries with the lowest GDP rate send the largest amount of their MSW to landfills, as landfilling is regarded the cheapest option in terms of investment (provided there are barely any regulatory government policies in place to divert waste from landfills). Yet, if the government introduces a high tax and landfilling fee, it may just be that it is more economically viable to reuse waste in order to produce energy than depositing it in landfills as the example of Estonia proves. Indeed, the most advanced country for averting its MSW from landfills is Estonia, which moved from landfilling almost all its MSW just after the Soviet era to only 5% in 2015. This is mainly due to energy recovery, which increased from 16% in 2012 up to 56% in 2014. However, the success in diverting waste from landfill to primarily energy recovery can potentially have a negative impact on separate collection and recycling schemes of MSW. The best recycling rate (61%) is demonstrated in Slovenia, especially in the Ljubljana area. This is due to its Regional Waste Management Centre (CERO Ljubljana), ýone of the most modern waste treatment facilities in Europe. Although Ljubljana in Slovenia and Tallinn in Estonia have been named as the best performers in terms of waste generation and recycling, the amount of waste generated by municipalities and waste management performance in individual municipalities differ considerably. The same experience can be shared in other countries, particularly in small municipalities, which as a rule of thumb, lack competence and resources to fulfil their waste management responsibilities, therefore, proving that further cooperation among municipalities and central government is essential.

While waste management and prevention policies are defined in all countries, a further focus to view waste as a source is lacking. Currently, Italy and the UK have the most WtE plants in terms of their capacity, with the aerobic and the aerobic/anaerobic treatments being successfully utilised in Italy, especially in the Milan area. The recent UK government encouragement to bring further focus on anaerobic digestion has proven also flourishing as the number of AD plants in the UK doubled during the two years after the Strategy was published. Yet, the UK exports relatively large amounts of MSW either for recycling, to long-distance destinations, such as China, India, and Turkey, or for energy recovery, mainly in Northern Europe, which can barely be justified from an environmental point of view (and does not comply with the principle of proximity). Even though Norway has 17 WtE plants, their capacity is small in contrast to its nearest neighbour Sweden (its waste export destination), which alongside with Denmark processes the largest amount of waste in Europe. It is important to note the current development of WtE plants in Central and Eastern Europe. Initially, WtE plants were considered a relatively expensive solution in the Baltic countries and Poland with landfill being the main viable option. Yet, this situation has changed – the WtE plants are now growing in the Baltic countries (except for Latvia) and Poland. During the past few years Poland has been able to build a system of WtE plants, which can produce 'green energy' through the effective utilisation of about 9% of MSW generated in the country with each newly built plant (five in total) being equipped with an efficient flue gas cleaning system to minimise any negative effects to the environment and human health. Estonia's experience has revealed that the capacity of WtE should be considered in advance to ensure that countries, especially with small markets, would have enough feedstock for their plants. Given that two modern WtE mass incineration plants are set to be built in Lithuania in the forthcoming years, it would be useful to see whether overcapacity will be avoided. Unlike Estonia, Lithuanian's NWMP categorically forbids any MSW import for energy recovery. Additionally, the near future will show whether (or not) the current State monopolies of the WtE industries in Estonia and Lithuania would lead to technology lock-in and, therefore, stagnation. It has been noted that flexibility in the choice of technology may improve the efficiency of WtE. At the other end of the spectrum, Greece and Latvia are two of the few European countries with no WtE infrastructure.

To conclude, it seems that most issues in the countries analysed are due to political factors and the decentralised nature of waste management with multi-level governance. WtE in the MSW management context varies significantly in the European countries considered. In most cases 'waste' is still regarded as a 'nuisance' rather than a resource. This must change when moving towards the circular economy. A political will is a key driver for change. Therefore, further cooperation between local waste management authorities and the different ministries responsible for diverse policies, including waste management, energy, and environment, should be employed in order to explore the full potential of the WtE industry embracing a holistic approach. There exists the CEWEP – the Confederation of European Waste-to-Energy Plants, the umbrella association of the owners and operators of WtE plants from 22 countries in Europe. Yet, it is essential for the industry to have regular links with the regulatory bodies, which could be facilitated via a network platform enabling a bottom-up approach. Building on the CEWEP and COOLSWEEP, a European-wide WtE network should be set up for different Member States (including the EEA) to share their experiences in research and development of innovative technology to unravel the potential in the WtE industry and any regulatory framework to accommodate it. Additionally, there is also a need for a top-down direction with the EU set targets being clearly communicated, as well as a future driven long-term long waste management strategy and priorities to achieve them being defined. Finally, further cooperation between Member States is largely advisable especially with neighbouring countries, as the example of Norway and Sweden indicates, where 'waste' could be transported to neighbouring countries with overcapacity for treatment, thus, benefiting both countries involved.

SEWAGE SLUDGE

Sewage sludge is the residual, semi-solid material that is produced as a by-product during sewage treatment of industrial or municipal wastewater. The term "septage" also refers to sludge from simple wastewater treatment but is connected to simple on-site sanitation systems, such as septic tanks.

When fresh sewage or wastewater enters a primary settling tank, approximately 50% of the suspended solid matter will settle out in an hour and a half. This collection of solids is known as raw sludge or primary solids and is said to be "fresh" before anaerobic processes become active. The sludge will become putrescent in a short time once anaerobic bacteria take over, and must be removed from the sedimentation tank before this happens.

This is accomplished in one of two ways. In an Imhoff tank, fresh sludge is passed through a slot to the lower story or digestion chamber where it is decomposed by anaerobic bacteria, resulting in liquefaction and reduced volume of the sludge. After digesting for an extended period, the result is called "digested" sludge and may be disposed of by drying and then landfilling. More commonly with domestic sewage, the fresh sludge is continuously extracted from the tank mechanically and passed to separate sludge digestion tanks that operate at higher temperatures than the lower story of the Imhoff tank and, as a result, digest much more rapidly and efficiently.

"Biosolids" is a term often used in conjunction with reuse of sewage solids after sewage sludge treatment. Biosolids can be defined as organic wastewater solids that can be reused after stabilization processes such as anaerobic digestion and composting. Opponents of sewage sludge reuse reject this term as a public relations term.

Classes of Treated Sewage Sludge

Class A sludge is typically dried and pasteurized, and is also known as "exceptional" quality.

Class B includes all sludge not classified as Class A. Class B sludge is typically "undigested" and is volatile.

Both classes of sludge may still contain radioactive or pharmaceutical wastes.

Quantities Produced

The amount of sewage sludge produced is proportional to the amount and concentration of wastewater treated, and it also depends on the type of wastewater treatment process used. It can be expressed as kg dry solids per cubic metre of wastewater treated. The total sludge production from a wastewater treatment process is the sum of sludge from primary settling tanks (if they are part of the process configuration) plus excess sludge from the biological treatment step. For example, primary sedimentation produces about 110–170 kg/ML of so-called primary sludge, with a value of 150 kg/ML regarded as being typical for municipal wastewater in the U.S. or Europe. The sludge production is expressed as kg of dry solids produced per ML of wastewater treated; one mega litre (ML) is 10^3 m^3. Of the biological treatment processes, the activated sludge process produces about 70–100 kg/ML of waste activated sludge, and a trickling filter process produces slightly less sludge from the biological part of the process: 60–100 kg/ML. This means that the total sludge production of an activated sludge process that uses primary sedimentation tanks is in the range of 180–270 kg/ML, being the sum of primary sludge and waste activated sludge.

United States municipal wastewater treatment plants in 1997 produced about 7.7 million dry tons of sewage sludge, and about 6.8 million dry tons in 1998 according to EPA estimates. As of 2004, about 60% of all sewage sludge was applied to land as a soil amendment and fertilizer for growing crops.

Production of sewage sludge can be reduced by conversion from flush toilets to dry toilets such urine-diverting dry toilets and composting toilets.

Contaminants

Pathogens

Bacteria in Class A sludge products can actually regrow under certain environmental conditions. Pathogens could easily remain undetected in untreated sewage sludge. Pathogens are not a significant health issue if sewage sludge is properly treated and site-specific management practices are followed.

Micro-pollutants

Micro-pollutants can become concentrated in sewage sludge. Each of these disposal options comes with myriad potential — and in some cases proven — human health and environment impacts.

Sterols and other hormones have been detected.

Heavy Metals

One of the main concerns in the treated sludge is the concentrated metals content (arsenic, cadmium, copper, etc., some of which are also critical plant micronutrients; certain metals are regulated while others are not. Leaching methods can be used to reduce the metal content and meet the regulatory limit.

In 2009 the EPA released the Targeted National Sewage Sludge Study, which reports on the level of metals, chemicals, hormones, and other materials present in a statistical sample of sewage sludges. Some highlights include:

- Silver is present to the degree of 20 mg/kg of sludge, on average, a near economically recoverable level, while some sludges of *exceptionally high quality* have up to 200 milligrams of silver per kilogram of sludge; one outlier demonstrated a silver lode of 800–900 mg per kg of sludge.
- Barium is present at the rate of 500 mg/kg, while manganese is present at the rate of 1 g/kg sludge.
- Lead, arsenic, chromium, and cadmium are estimated by the EPA to be present in detectable quantities in 100% of national sewage sludges in the US, while thallium is only estimated to be present in 94.1% of sludges.

Other Hazardous Substances

Sewage treatment plants receive various forms of hazardous waste from hospitals, nursing homes, industry and households. Low levels of constituents such as PCBs, dioxin, and brominated flame retardants, may remain in treated sludge. There are potentially thousands other components of sludge that remain untested/undetected disposed of from modern society that also end up in sludge (pharmaceuticals, nano particles, etc.) which have been proven to be hazardous to both human and ecological health.

In 2013 in South Carolina PCBs were discovered in very high levels in wastewater sludge. The problem was not discovered until thousands of acres of farm land in South Carolina were contaminated by this hazardous material. SCDHEC issued emergency regulatory order banning all PCB laden sewage sludge from being land applied on farm fields or deposited into landfills in South Carolina.

Also in 2013, after DHEC request, the city of Charlotte decided to stop land applying sewage sludge in South Carolina while authorities investigated the source of PCB contamination. In February 2014, the city of Charlotte admitted PCBs have entered their sewage treatment centers as well.

Contaminants of concern in sewage sludge are plasticizers, PDBEs, and others generated by human activities, including personal care products and medicines. Synthetic fibers from fabrics persist in treated sewage sludge as well as in biosolids-treated soils and may thus serve as an indicator of past biosolids application.

Pollutant Ceiling Concentration

The term "pollutant" is defined as part of the EPA 503 rule. The components of sludge have pollutant limits defined by the EPA. "A Pollutant is an organic substance, an inorganic substance, a combination of organic and inorganic substances, or a pathogenic organism that, after discharge and upon exposure, ingestion, inhalation, or assimilation into an organism either directly from the environment or indirectly by ingestion through the food chain, could, on the basis of information available to the Administrator of EPA, cause death, disease, behavioral abnormalities, cancer, genetic mutations, physiological malfunctions (including malfunction in reproduction), or physical deformations in either organisms or offspring of the organisms." The maximum component pollutant limits by the US EPA are:

Pollutant	Ceiling concentration (milligrams per kilogram)
Cadmium	85
Copper	4300
Lead	840
Mercury	57

Molybdenum	75
Nickel	420
Selenium	100
Zinc	7500

Treatment

Sewage sludge treatment is the process of removing contaminants from wastewater. Sewage sludge is produced from the treatment of wastewater in sewage treatment plants and consists of two basic forms — raw primary sludge and secondary sludge, also known as activated sludge in the case of the activated sludge process.

Sewage sludge is usually treated by one or several of the following treatment steps: lime stabilization, thickening, dewatering, drying, anaerobic digestion or composting. Some treatment processes, such as composting and alkaline stabilization, that involve significant amendments may affect contaminant strength and concentration: depending on the process and the contaminant in question, treatment may decrease or in some cases increase the bioavailability and/or solubility of contaminants.

Following treatment, sewage sludge is either landfilled, incinerated, applied on agricultural land or, in some cases, retailed or given away for free to the general public.

Health Risks

In 2011, the EPA commissioned a study at the United States National Research Council (NRC) to determine the health risks of sludge. In this document the NRC pointed out that many of the dangers of sludge are unknown and unassessed. Additionally "Regulations that limit contact with biosolids do not prevent environmental processes in the conceptual model such as aerosolization or erosion and the death or multiplication of pathogens."

The National Research Council published "Biosolids Applied to Land: Advancing Standards and Practices" in July 2002. The NRC concluded that while there is no documented scientific evidence that sewage sludge regulations have failed to protect public health, there is persistent uncertainty on possible adverse health effects. The NRC noted that further research is needed and made about 60 recommendations for addressing public health concerns, scientific uncertainties, and data gaps in the science underlying the sewage sludge standards. The EPA responded with a commitment to conduct research addressing the NRC recommendations.

Residents living near Class B sludge processing sites may experience asthma or pulmonary distress due to bioaerosols released from sludge fields.

A 2004 survey of 48 individuals near affected sites found that most reported irritation symptoms, about half reported an infection within a month of the application, and about a fourth were affected by *Staphylococcus aureus*, including two deaths. The number of reported S. aureus infections was 25 times as high as in hospitalized patients, a high-risk group. The authors point out that regulations call for protective gear when handling Class B biosolids and that similar protections could be considered for residents in nearby areas given the wind conditions.

Khuder, Milz, Bisesi, Vincent, McNulty, and Czajkowski conducted a health survey of persons living in close proximity to Class B sludged land. A sample of 437 people exposed to Class B sludge (living within 1-mile (1.6 km) of sludged land) - and using a control group of 176 people not exposed to sludge (not living within 1-mile (1.6 km) of sludged land) reported the following:

> "Results revealed that some reported health-related symptoms were statistically significantly elevated among the exposed residents, including excessive secretion of tears, abdominal bloating, jaundice, skin ulcer, dehydration, weight loss, and general weakness. The frequency of reported occurrence of bronchitis, upper respiratory infection, and giardiasis were also statistically significantly elevated. The findings suggest an increased risk for certain respiratory, gastrointestinal, and other diseases among residents living near farm fields on which the use of biosolids was permitted".
>
> — Khuder et al., Health Survey of Residents Living near Farm Fields Permitted to Receive Biosolids

Although correlation does not imply causation, such extensive correlations may lead reasonable people to conclude that precaution is necessary in dealing with sludge and sludged farmlands.

Harrison and Oakes suggest that, in particular, "until investigations are carried out that answer these questions (...about the safety of Class B sludge...), land application of Class B sludges should be viewed as a practice that subjects neighbors and workers to substantial risk of disease." They further suggest that even Class A treated sludge may have chemical contaminants (including heavy metals, such as lead) or endotoxins present, and a precautionary approach may be justified on this basis, though the vast majority of incidents reported by Lewis, et al. have been correlated with exposure to Class B untreated sludge and not Class A treated sludge.

A 2005 report by the state of North Carolina concluded that "a surveillance program of humans living near application sites should be developed to determine if there are adverse health effects in humans and animals as a result of biosolids application."

Disposal

After treatment, and dependent upon the quality of sludge produced (for example with regards to heavy metal content), sewage sludge is most commonly either disposed of in landfills, dumped in the ocean or applied to land for its fertilizing properties, as pioneered by the product Milorganite.

Landfill

Sewage sludge deposition in landfills can circulate human-virulent species of Cryptosporidium and Guardia pathogens. Sonication and quicklime stabilization are most effective in inactivation of these pathogens; microwave energy disintegration and topsoil stabilization were less effective.

Ocean Dumping

It used to be common practice to dump sewage sludge into the ocean, however, this practice has stopped in many nations due to environmental concerns as well to domestic and international laws and treaties.

Land Application

Biosolids is a term widely used to denote the byproduct of domestic and commercial sewage and wastewater treatment that is to be used in agriculture. National regulations that dictate the practice of land application of treated sewage sludge differ widely and e.g. in the US there are widespread disputes about this practice.

Depending on their level of treatment and resultant pollutant content, biosolids can be used in regulated applications for non-food agriculture, food agriculture, or distribution for unlimited use. Treated biosolids can be produced in cake, granular, pellet, or liquid form and are spread over land before being incorporated into the soil or injected directly into the soil by specialist contractors. Such use was pioneered by the production of Milorganite in 1926.

Use of sewage sludge has shown an increase in level of soil available phosphorus and soil salinity. The findings of a 20-year field study of air, land, and water in Arizona, concluded that use of biosolids is sustainable and improves the soil and crops. Other studies report that plants uptake large quantities of heavy metals and toxic pollutants that are retained by produce, which is then consumed by humans.

A PhD thesis studying the addition of sludge to neutralize soil acidity concluded that the practice was not recommended if large amounts are used because the sludge produces acids when it oxidizes.

Studies have indicated that pharmaceuticals and personal care products, which often adsorb to sludge during wastewater treatment, can persist in agricultural soils following biosolid application. Some of these chemicals, including potential endocrine disruptor Triclosan, can also travel through the soil column and leach into agricultural tile drainage at detectable levels. Other studies, however, have shown that these chemicals remain adsorbed to surface soil particles, making them more susceptible to surface erosion than infiltration. These studies are also mixed in their findings regarding the persistence of chemicals such as triclosan, triclocarban, and other pharmaceuticals. The impact of this persistence in soils is unknown, but the link to human and land animal health is likely tied to the capacity for plants to absorb and accumulate these chemicals in their consumed tissues. Studies of this kind are in early stages, but evidence of root uptake and translocation to leaves did occur for both triclosan and triclocarban in soybeans. This effect was not present in corn when tested in a different study.

A cautionary approach to land application of biosolids has been advocated by some for regions where soils have lower capacities for toxics absorption or due to the presence of unknowns in sewage biosolids. In 2007 the Northeast Regional Multi-State Research Committee (NEC 1001) issued conservative guidelines tailored to the soils and conditions typical of the northeastern US.

Use of sewage sludge is prohibited for produce to be labeled USDA-certified organic. In 2014 the United States grocery chain Whole Foods banned produce grown in sewage sludge.

Treated sewage sludge has been used in the UK, Europe and China agriculturally for more than 80 years, though there is increasing pressure in some countries to stop the practice of land application due to farm land contamination and public outrage. In the 1990s there was pressure in some European countries to ban the use of sewage sludge as a fertilizer. Switzerland, Sweden, Austria, and others introduced a ban. Since the 1960s there has been cooperative activity with industry to reduce the inputs of persistent substances from factories. This has been very successful and, for example, the content of cadmium in sewage sludge in major European cities is now only 1% of what it was in 1970.

Incineration

Sludge can also be incinerated in sludge incineration plants which comes with its own set of environmental concerns (air pollution, disposal of the ash). Pyrolysis of the sludge to create syngas and potentially biochar is possible, as is combustion of biofuel produced from drying sewage sludge or incineration in a waste-to-energy facility for direct production of electricity and steam for district heating or industrial uses.

Thermal processes can greatly reduce the volume of the sludge, as well as achieve remediation of all or some of the biological concerns. Direct waste-to-energy incineration and complete combustion systems (such as the Gate 5 Energy System) will require multi-step cleaning of the exhaust gas, to ensure no hazardous substances are released. In addition, the ash produced by incineration or incomplete combustion processes (such as fluidized-bed dryers) may be difficult to use without subsequent treatment due to high heavy metal content; solutions to this include leaching of the ashes to remove heavy metals or in the case of ash produced in a complete-combustion process, or with biochar produced from a pyrolytic process, the heavy metals may be fixed in place and the ash material readily usable as a LEEDs preferred additive to concrete or asphalt. Examples of other ways to use dried sewage sludge as an energy resource include the Gate 5 Energy System, an innovative process to power a steam turbine using heat from burning milled and dried sewage sludge, or combining dried sewage sludge with coal in coal-fired power stations. In both cases this allows for production of electricity with less carbon-dioxide emissions than conventional coal-fired power stations.

References

- Biomass-feedstocks, bioenergy: energy.gov, Retrieved 20 July, 2019
- Karimi alavijeh, masih; yaghmaei, soheila. "biochemical production of bioenergy from agricultural crops and residue in iran". Waste management. 52: 375–394. Doi:10.1016/j.wasman.2016.03.025
- Ibss-feedstock-characteristics-1511: ces.ncsu.edu, Retrieved 15 May, 2019
- "Jatropha curcas". Germplasm resources information network (grin). Agricultural research service (ars), united states department of agriculture(usda). Retrieved 2010-10-14
- Janick, jules; robert e. Paull (2008). The encyclopedia of fruit & nuts. Cabi. Pp. 371–372. Isbn 978-0-85199-638-7
- What-is-lignocellulose: bioenergyconsult.com, Retrieved 23 June, 2019
- Tchobanoglous, george; burton, franklin l.; stensel, h. David (2003). Wastewater engineering : treatment and reuse (4 ed.). Metcalf & eddy. P. 1449. Isbn 978-0071122504
- "Pharmaceutical waste management". Premier. Archived from the originalon 25 may 2007. Retrieved 30 may 2017

Biomass Conversion Techniques

CHAPTER 4

Some of the techniques used for the conversion of biomass are thermal conversion, bio-conversion, chemical conversion, biochemical conversion, thermal depolymerization and electrochemical conversion. The diverse applications of these biomass conversion techniques have been thoroughly discussed in this chapter.

THERMAL CONVERSION OF BIOMASS

A wide range of technologies exists to convert the energy stored in biomass to more useful forms of energy. These technologies can be classified according to the principal energy carrier produced in the conversion process. Carriers are in the form of heat, gas, liquid and/or solid products, depending on the extent to which oxygen is admitted to the conversion process (usually as air). The major methods of thermal conversion are combustion in excess air, gasification in reduced air, and pyrolysis in the absence of air.

Combustion

Conventional combustion technologies raise steam through the combustion of biomass. This steam may then be expanded through a conventional turbo-alternator to produce electricity. A number of combustion technology variants have been developed. Underfeed stokers are suitable for small scale boilers up to 6 MWth. Grate type boilers are widely deployed. They have relatively low investment costs, low operating costs and good operation at partial loads. However, they can have higher NOx emissions and decreased efficiencies due to the requirement of excess air, and they have lower efficiencies.

Fluidized bed combustors (FBC), which use a bed of hot inert material such as sand, are a more recent development. Bubbling FBCs are generally used at 10-30 MWth capacity, while Circulating FBCs are more applicable at larger scales. Advantages of FBCs are that they can tolerate a wider range of poor quality fuel, while emitting lower NOx levels.

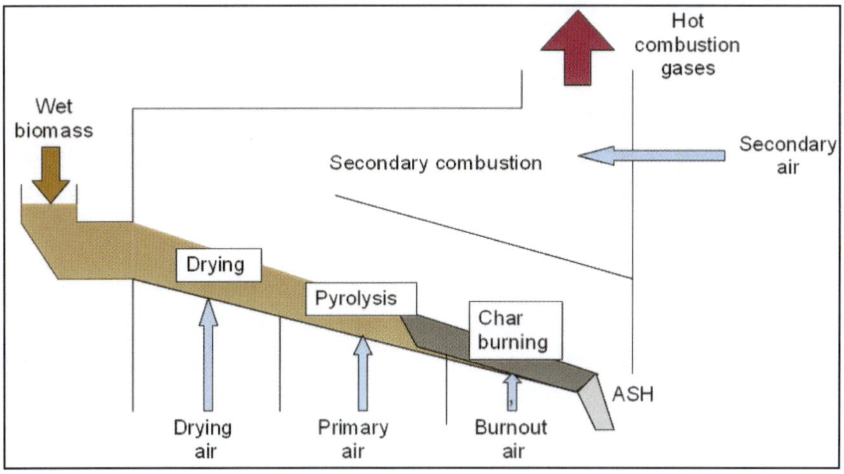

Co-firing

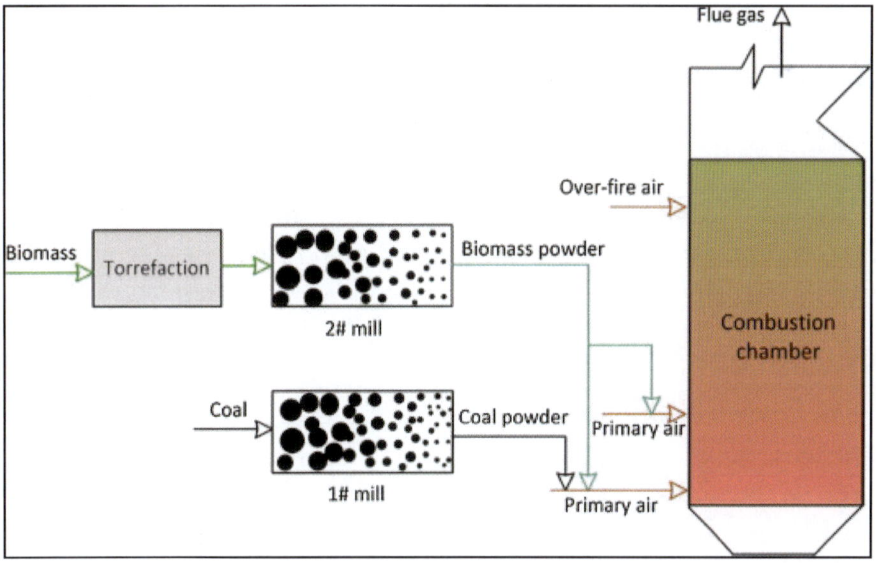

Co-firing or co-combustion of biomass wastes with coal and other fossil fuels can provide a short-term, low-risk, low-cost option for producing renewable energy while simultaneously reducing the use of fossil fuels. Co-firing involves utilizing existing power generating plants that are fired with fossil fuel (generally coal), and displacing a small proportion of the fossil fuel with renewable biomass fuels. Co-firing has the major advantage of avoiding the construction of new, dedicated, waste-to-energy power plant. Co-firing may be implemented using different types and percentages of wastes in a range of combustion and gasification technologies. Most forms of biomass wastes are suitable for co-firing. These include dedicated municipal solid wastes, wood waste and agricultural residues such as straw and husk.

Pyrolysis

Pyrolysis is the thermal decomposition of materials at elevated temperatures in an inert atmosphere. It involves a change of chemical composition and is irreversible. It is most commonly used in the treatment of organic materials. It is one of the processes involved in charring wood. In general, pyrolysis of organic substances produces volatile products and leaves a solid residue enriched in carbon, char. Extreme pyrolysis, which leaves mostly carbon as the residue, is called carbonization.

The process is used heavily in the chemical industry, for example, to produce ethylene, many forms of carbon, and other chemicals from petroleum, coal, and even wood, to produce coke from coal. Aspirational applications of pyrolysis would convert biomass into syngas and biochar, waste plastics back into usable oil, or waste into safely disposable substances.

Pyrolysis is one of various types of chemical degradation processes that occur at higher temperatures (above the boiling point of water or other solvents). It differs from other processes like combustion and hydrolysis in that it usually does not involve the addition of other reagents such as oxygen (O_2, in combustion) or water (in hydrolysis). Pyrolysis produces solids (char), condensable liquids (tar), and uncondensing/permanent gasses.

Types of Pyrolysis

Complete pyrolysis of organic matter usually leaves a solid residue that consists mostly of elemental carbon; the process is then called carbonization. More specific cases of pyrolysis include:

- Dry distillation, as in the original production of sulfuric acid from sulfates.
- Destructive distillation, as in the manufacture of charcoal, coke and activated carbon.
- Caramelization of sugars.
- High-temperature cooking processes such as roasting, frying, toasting, and grilling.
- Charcoal burning, the production of charcoal.
- Tar production by pyrolysis of wood in tar kilns.

- Cracking of heavier hydrocarbons into lighter ones, as in oil refining.
- Thermal depolymerization, that breaks down plastics and other polymers into monomers and oligomers.
- Hydrous pyrolysis, in the presence of superheated water or steam, also used in oil refining.
- Ceramization involving the formation of polymer derived ceramics from preceramic polymers under an inert atmosphere.
- Catagenesis, the natural conversion of buried organic matter to fossil fuels.
- Flash vacuum pyrolysis, used in organic synthesis.

General Processes and Mechanisms

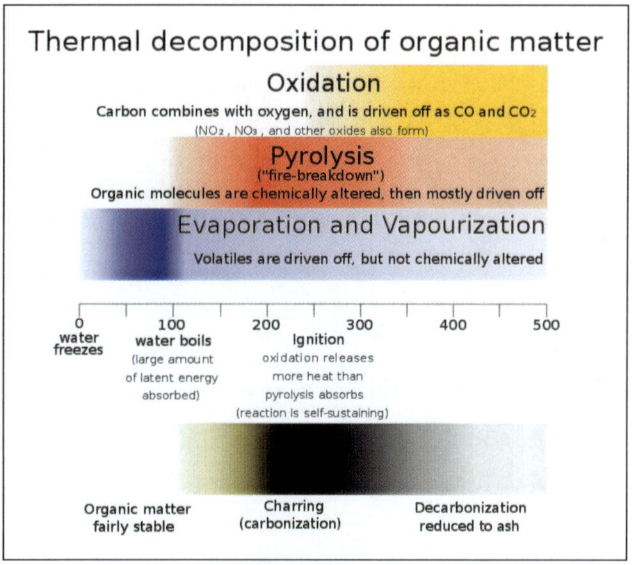

Processes in the thermal degradation of organic matter at atmospheric pressure.

Pyrolysis generally consists in heating the material above its decomposition temperature, breaking chemical bonds in its molecules. The fragments usually become smaller molecules, but may combine to produce residues with larger molecular mass, even amorphous covalent solids.

In many settings, some amounts of oxygen, water, or other substances may be present, so that combustion, hydrolysis, or other chemical processes may occur besides pyrolysis proper. Sometimes those chemical are added intentionally, as in the burning of firewood, in the traditional manufacture of charcoal, and in the steam cracking of crude oil.

Conversely, the starting material may be heated in a vacuum or in an inert atmosphere to avoid adverse chemical reactions. Pyrolysis in a vacuum also lowers the boiling point of the byproducts, improving their recovery.

Burning pieces of wood, showing various stages of pyrolysis followed by oxidative combustion.

When organic matter is heated at increasing temperatures in open containers, the following processes generally occur, in successive or overlapping stages:

- Below about 100 °C, volatiles, including some water, evaporate. Heat-sensitive substances, such as vitamin C and proteins, may partially change or decompose already at this stage.
- At about 100 °C or slightly higher, any remaining water that is merely absorbed in the material is driven off. Water trapped in crystal structure of hydrates may come off at somewhat higher temperatures. This process consumes a lot of energy, so the temperature may stop rising until this stage is complete.
- Some solid substances, like fats, waxes, and sugars, may melt and separate.
- Between 100 and 500 °C, many common organic molecules break down. Most sugars start decomposing at 160-180 °C. Cellulose, a major component of wood, paper, and cotton fabrics, decomposes at about 350 °C. Lignin, another major wood component, starts decomposing at about 350 °C, but continues releasing volatile products up to 500 °C. The decomposition products usually include water, carbon monoxide CO_2 and/or carbon dioxide CO_2 2, as well as a large number of organic compounds. Gases and volatile products leave the sample, and some of them may condense again as smoke. Generally, this process also absorbs energy. Some volatiles may ignite and burn, creating a visible flame. The non-volatile residues typically become richer in carbon and form large disordered molecules, with colors ranging between brown and black. At this point the matter is said to have been "charred" or "carbonized".
- At 200-300 °C, if oxygen has not been excluded, the carbonaceous residue may start to burn, in a highly exothermic reaction, often with no or little visible flame. Once carbon combustion starts, the temperature rises spontaneously, turning the residue into a glowing ember and releasing carbon dioxide and/or monoxide. At this stage, some of the nitrogen still remaining in the residue may

be oxidized into nitrogen oxides like NO_2 and N_2O_3. Sulfur and other elements like chlorine and arsenic may be oxidized and volatilized at this stage.

- Once combustion of the carbonaceous residue is complete, a powdery or solid mineral residue (ash) is often left behind, consisting of inorganic oxidized materials of high melting point. Some of the ash may have left during combustion, entrained by the gases as fly ash or particulate emissions. Metals present in the original matter usually remain in the ash as oxides or carbonates, such as potash. Phosphorus, from materials such as bone, phospholipids, and nucleic acids, usually remains as phosphates.

Occurrence and Uses

Cooking

Caramelized onions are slightly pyrolyzed.

Pyrolysis has many applications in food preparation. Caramelization is the pyrolysis of sugars in food (often after the sugars have been produced by the breakdown of polysaccharides). The food goes brown and changes flavour. The distinctive flavours are used in many dishes; for instance, caramelized onion is used in French onion soup. The temperatures needed for caremelization lie above the boiling point of water. Frying oil can easily rise above boiling point. Putting a lid on the frying pan keeps the water in, and some of it re-condenses, keeping the temperature too cool to brown for longer.

This pizza is pyrolyzed, possibly almost completely carbonized.

Pyrolysis of food can also be undesirable, as in the charring of burnt food (at temperatures too low for the oxidative combustion of carbon to produce flames and burn the food to ash).

Coke, Carbon, Charcoals and Chars

Charcoal briquettes, often made from compressed sawdust or similar, in use.

Carbon and carbon-rich materials have desirable properties but are nonvolatile, even at high temperatures. Consequently, pyrolysis is used to produce many kinds of carbon; these can be used for fuel, as reagents in steelmaking (coke), and as structural materials.

Charcoal is a less smoky fuel than unpyrolized wood). Some cities ban, or used to ban, wood fires; when residents only use charcoal (and similarly-treated rock coal, called *coke*) air pollution is significantly reduced. In cities where people do not generally cook or heat with fires, this is not needed. In the mid-20th century, "smokeless" legislation in Europe required cleaner-burning techniques, such as coke fuel and smoke-burning incinerators as an effective measure to reduce air pollution.

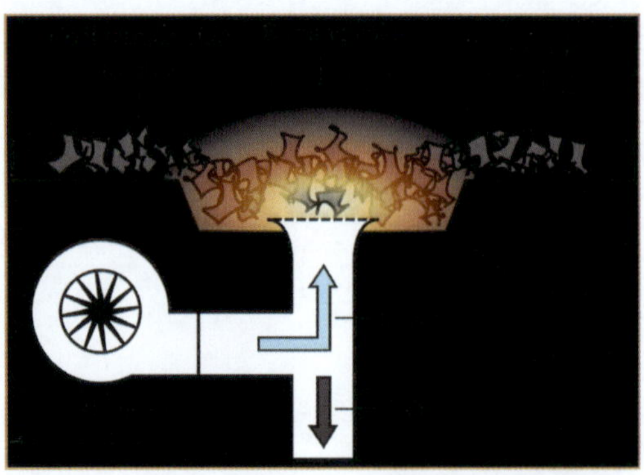

A blacksmith's forge, with a blower forcing air through a bed of fuel to raise the temperature of the fire. On the periphery, coal is pyrolyized, absorbing heat; the coke at the center is almost pure carbon, and releases a lot of heat when the carbon oxidizes.

The coke-making or "coking" process consists of heating the material in "coking ovens" to very high temperatures (up to 900 °C or 1,700 °F) so that those molecules are broken down into lighter volatile substances, which leave the vessel, and a porous but hard residue that is mostly carbon and inorganic ash. The amount of volatiles varies with the source material, but is typically 25–30% of it by weight. High temperature pyrolysis is used on an industrial scale to convert coal into coke. This is useful in metallurgy, where the higher temperatures are necessary for many processes, such as steelmaking. Volatile by-products of this process are also often useful, including benzene and pyridine. Coke can also be produced from the solid residue left from petroleum refining.

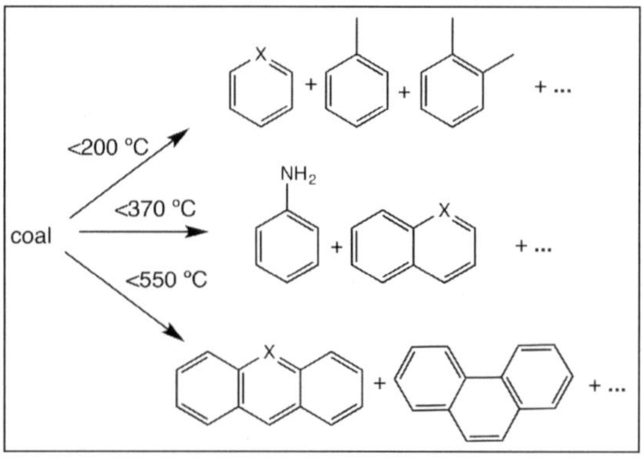

Typical organic products obtained by pyrolysis of coal (X = CH, N).

The original vascular structure of the wood and the pores created by escaping gases combine to produce a light and porous material. By starting with a dense wood-like material, such as nutshells or peach stones, one obtains a form of charcoal with particularly fine pores (and hence a much larger pore surface area), called activated carbon, which is used as an adsorbent for a wide range of chemical substances.

Biochar is the residue of incomplete organic pyrolysis, e.g., from cooking fires. They are a key component of the terra preta soils associated with ancient indigenous communities of the Amazon basin. Terra preta is much sought by local farmers for its superior fertility and capacity to promote and retain an enhanced suite of beneficial microbiota, compared to the typical red soil of the region. Efforts are underway to recreate these soils through biochar, the solid residue of pyrolysis of various materials, mostly organic waste.

Carbon fibers are filaments of carbon that can be used to make very strong yarns and textiles. Carbon fiber items are often produced by spinning and weaving the desired item from fibers of a suitable polymer, and then pyrolyzing the material at a high temperature (from 1,500–3,000 °C or 2,730–5,430 °F). The first carbon fibers were made from rayon, but polyacrylonitrile has become the most common starting material. For their first workable electric lamps, Joseph Wilson Swan and Thomas Edison used carbon filaments made by pyrolysis of cotton yarns and bamboo splinters, respectively.

Pyrolysis is the reaction used to coat a preformed substrate with a layer of pyrolytic carbon. This is typically done in a fluidized bed reactor heated to 1,000–2,000 °C or 1,830–3,630 °F. Pyrolytic carbon coatings are used in many applications, including artificial heart valves.

Liquid and Gaseous Biofuels

Pyrolysis is the basis of several methods for producing fuel from biomass, i.e. lignocellulosic biomass. Crops studied as biomass feedstock for pyrolysis include native North American prairie grasses such as *switchgrass* and bred versions of other grasses such as *Miscantheus giganteus*. Other sources of organic matter as feedstock for pyrolysis include greenwaste, sawdust, waste wood, nut shells, straw, cotton trash, rice hulls. Animal waste including poultry litter, dairy manure, and potentially other manures are also under evaluation. Some industrial byproducts are also suitable feedstock including paper sludge and distillers grain.

Synthetic diesel fuel by pyrolysis of organic materials is not yet economically competitive. Higher efficiency is sometimes achieved by flash pyrolysis, in which finely divided feedstock is quickly heated to between 350 and 500 °C (660 and 930 °F) for less than two seconds.

Syngas is usually Produced by Pyrolysis. The low quality of oils produced through pyrolysis can be improved by physical and chemical processes, which might drive up production costs, but may make sense economically as circumstances change.

There is also the possibility of integrating with other processes such as mechanical biological treatment and anaerobic digestion. Fast pyrolysis is also investigated for biomass conversions. Fuel bio-oil can also be produced by hydrous pyrolysis.

Ethylene

Pyrolysis is used to produce ethylene, the chemical compound produced on the largest scale industrially (>110 million tons/year in 2005). In this process, hydrocarbons from petroleum are heated to around 600 °C (1,112 °F) in the presence of steam; this is called steam cracking. The resulting ethylene is used to make antifreeze (ethylene glycol), PVC (via vinyl chloride), and many other polymers, such as polyethylene and polystyrene.

Semiconductors

The process of metalorganic vapour phase epitaxy (MOCVD) entails pyrolysis of volatile organometallic compounds to give semiconductors, hard coatings, and other applicable materials. The reactions entail thermal degradation of precursors, with deposition of the inorganic component and release of the hydrocarbons as gaseous waste. Since it is an atom-by-atom deposition, these atoms organize themselves into crystals to form the bulk semiconductor. Silicon chips are produced by the pyrolysis of silane:

$$SiH_4 \rightarrow Si + 2\,H_2$$

Gallium arsenide, another semiconductor, forms upon co-pyrolysis of trimethylgallium and arsine.

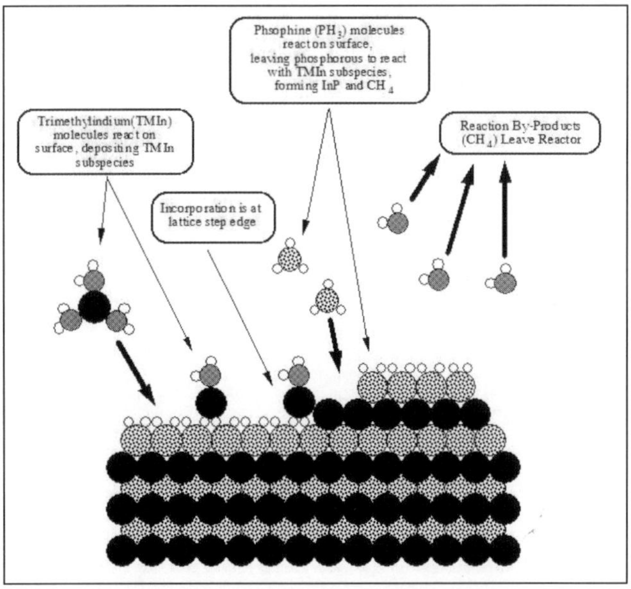

Illustration of the MOCVD process, which entails pyrolysis of volatiles.

Recycling

Pyrolysis can also be used to treat plastic waste. The main advantage is the reduction in volume of the waste. In principle, pyrolysis will regenerate the monomers (precursors) to the polymers that are treated, but in practice the process is neither a clean nor an economically competitive source of monomers.

In tire recycling, tire pyrolysis is well developed technology. Other products from car tire pyrolysis include steel wires, carbon black and bitumen. The area faces legislative, economic, and marketing obstacles. Oil derived from tire rubber pyrolysis contains high sulfur content, which gives it high potential as a pollutant and should be desulfurized.

Thermal Cleaning

Pyrolysis is also used for *thermal cleaning*, an industrial application to remove organic substances such as polymers, plastics and coatings from parts, products or production components like extruder screws, spinnerets and static mixers. During the thermal cleaning process, at temperatures between 310 °C to 540 °C (600 °F to 1000 °F), organic material is converted by pyrolysis and oxidation into volatile organic compounds, hydrocarbons and carbonized gas. Inorganic elements remain.

Several types of thermal cleaning systems use pyrolysis:

- Molten Salt Baths belong to the oldest thermal cleaning systems; cleaning with a molten salt bath is very fast but implies the risk of dangerous splatters, or other potential hazards connected with the use of salt baths, like explosions or highly toxic hydrogen cyanide gas.

- Fluidized Bed Systems use sand or aluminium oxide as heating medium; these systems also clean very fast but the medium does not melt or boil, nor emit any vapors or odors; the cleaning process takes one to two hours.
- Vacuum Ovens use pyrolysis in a vacuum avoiding uncontrolled combustion inside the cleaning chamber; the cleaning process takes 8 to 30 hours.
- Burn-Off Ovens, also known as Heat-Cleaning Ovens, are gas-fired and used in the painting, coatings, electric motors and plastics industries for removing organics from heavy and large metal parts.

Fine Chemical Synthesis

Pyrolysis is used in the production of chemical compounds, mainly, but not only, in the research laboratory.

The area of boron-hydride clusters started with the study of the pyrolysis of diborane (B_2H_6) at ca. 200 °C. Products include the clusters pentaborane and decaborane. These pyrolyses involve not only cracking (to give H_2), but also recondensation.

The synthesis of nanoparticles, zirconia and oxides utilizing an ultrasonic nozzle in a process called ultrasonic spray pyrolysis (USP).

Other Uses and Occurrences

Pyrolysis is used to turn organic materials into carbon for the purpose of carbon-14 dating.

- Py-GC-MS is an important laboratory procedure to determine the structure of compounds.
- Pyrolysis of tobacco, paper, and additives, in cigarettes and other products, generates many volatile products (including nicotine, carbon monoxide, and tar) that are responsible for the aroma and health effects of smoking. Similar considerations apply to the smoking of marijuana and the burning of incense products and mosquito coils.
- Pyrolysis occurs during the incineration of trash, potentially generating volatiles that are toxic or contribute to air pollution if not completely burned.
- Laboratory or industrial equipment sometimes gets fouled by carbonaceous residues that result from coking, the pyrolysis of organic products that come into contact with hot surfaces.

Torrefaction

Torrefaction of biomass, e.g., wood or grain, is a mild form of pyrolysis at temperatures typically between 200 and 320 °C. Torrefaction changes biomass properties to provide a better fuel quality for combustion and gasification applications. Torrefaction produces a dry product with no biological activity like rotting. Torrefaction combined with densification creates an energy-dense fuel carrier of 20 to 21 GJ/ton lower heating value (LHV). Torrefaction makes the material undergo Maillard reactions.

Biomass can be an important energy source. However, nature provides a large diversity of biomass with varying characteristics. To create highly efficient biomass-to-energy chains, torrefaction of biomass in combination with densification (pelletisation or briquetting) is a promising step to overcome logistic economics in large-scale sustainable energy solutions, i.e. make it easier to transport and store it. Pellets or briquets are lighter, drier, and more stable in storage than the biomass they derive from.

Process

Torrefaction is a thermochemical treatment of biomass at 200 to 320 °C (392 to 608 °F). It is carried out under atmospheric pressure and in the absence of oxygen, i.e. with no air. During the torrefaction process, the water contained in the biomass as well as superfluous volatiles are released, and the biopolymers (cellulose, hemicellulose and lignin) partly decompose, giving off various types of volatiles. The final product is the remaining solid, dry, blackened material that is referred to as *torrefied biomass* or *bio-coal.*

During the process, the biomass typically loses 20% of its mass (bone dry basis) and 10% of its heating value, with no appreciable change in volume. This energy (the volatiles) can be used as a heating fuel for the torrefaction process. After the biomass is torrefied it can be densified, usually into briquettes or pellets using conventional densification equipment, to increase its mass and energy density and to improve its hydrophobic properties. The final product may repel water and thus can be stored in moist air or rain without appreciable change in moisture content or heating value, unlike the original biomass.

The history of torrefaction dates to the beginning of the 19th century, and gasifiers were used on a large scale during the Second World War.

Added Value of Torrefied Biomass

Torrefied and densified biomass has several advantages in different markets, which makes it a competitive option compared to conventional biomass wood pellets.

- Higher energy density:

  An energy density of 18–20 GJ/m³ — compared to the 26 to 33 gigajoules per tonne heat content of natural anthracite coal — can be achieved when combined with densification (pelletizing or briquetting) compared to values of 10–11 GJ/m³ for raw biomass, driving a 40–50% reduction in transportation costs. Importantly, pelletizing or briquetting primarily increases energy density. Torrefaction alone typically decreases energy density, though it makes the material easier to make into pellets or briquettes.

- More homogeneous composition:

  Torrefied biomass can be produced from a wide variety of raw biomass feedstocks that yield similar product properties. Most woody and herbaceous biomass consists of three main polymeric structures: cellulose, hemicellulose and lignin. Together these are called lignocellulose. Torrefaction primarily drives moisture and oxygen-rich and hydrogen-rich functional groups from these structures, producing similar char-like structures in all three cases. Therefore, most biomass fuels, regardless of origin, produce torrefied products with similar properties—with the exception of ash properties, which largely reflect the original fuel ash content and composition.

- Hydrophobic behavior:

  Torrefied biomass has hydrophobic properties, i.e., repels water, and when combined with densification make bulk storage in open air feasible.

- Elimination of biological activity:

  All biological activity is stopped, reducing the risk of fire and stopping biological decomposition like rotting.

- Improved grindability:

  Torrefaction of biomass leads to improved grindability of biomass. This leads to more efficient co-firing in existing coal-fired power stations or entrained-flow gasification for the production of chemicals and transportation fuels.

Markets for Torrefied Biomass

Torrefied biomass has added value for different markets. Biomass in general provides a low-cost, low-risk route to lower CO_2-emissions. When high volumes are needed, torrefaction can make biomass from distant sources price competitive because the denser material is easier to store and transport.

- Wood powder fuel:
  - Torrefied wood powder can be ground into a fine powder and when compressed, mimics liquefied petroleum gas (LPG).
- Large-scale co-firing in coal-fired power plants:
  - Torrefied biomass results in lower handling costs.

  - Torrefied biomass enables higher co-firing rates.
  - Product can be delivered in a range of LHVs (20–25 GJ/ton) and sizes (briquette, pellet).
  - Co-firing torrefied biomass with coal leads to reduction in net power plant emissions.
- Steel production:
  - Fibrous biomass is very difficult to deploy in furnaces.
  - To replace injection coal, biomass product needs to have LHV of more than 25 GJ/ton.
- Residential/decentralized heating:
  - Relatively high percentage of transport on wheels in the supply chain makes biomass expensive. Increasing volumetric energy density does decrease costs.
  - Limited storage space increases need for increased volumetric density.
  - Moisture content important as moisture leads to smoke and smell.
- Biomass-to-Liquids:
  - Torrefied biomass results in lower handling costs.
  - Torrefied biomass serves as a 'clean' feedstock for production of transportation fuels (Fischer–Tropsch process), which saves on production costs.
- Miscellaneous uses:
  - Several guitar builders have used torrefaction to obtain more dimensionally stable wood for guitar parts than traditional kiln-drying or air-drying provides, including Martin, Gibson, and luthier Dana Bourgeois.

Gasification

Gasification is a technology that converts carbon-containing materials, including coal, waste and biomass, into synthetic gas which in turn can be used to produce electricity and other valuable products, such as chemicals, fuels, and fertilizers.

Gasification does not involve combustion, but instead uses little or no oxygen or air in a closed reactor to convert carbon-based materials directly into a synthetic gas, or syngas.

Gasification can recover the energy locked in biomass and municipal solid waste, converting those materials into valuable products and eliminating the need for incineration or landfilling. Metals and glass should be segregated from the waste stream prior to being sent into the gasification process. The gasification process breaks these carbon-containing materials down to the molecular level, so impurities like nitrogen, sulphur, and mercury can be easily removed and sold as valuable industrial commodities.

While syngas is the primary product of the gasification plants, marketable products obtained from syngas include chemicals (45%), liquid fuels (28%), gaseous fuels (8%), and electric power (19%).

In some coal combustion-based power plants, only a third of the energy value of coal is actually converted into electricity. A coal gasification power plant, however, typically gets dual duty from the gases it produces. First, the coal gases, cleaned of impurities, are fired in a gas turbine - much like natural gas - to generate one source of electricity. The hot exhaust of the gas turbine, and some of the heat generated in the gasification process, are then used to generate steam for use in a steam turbine-generator. This dual source of electric power, called a "combined cycle," is much more efficient in converting coal's energy into usable electricity. The fuel efficiency of a coal gasification power plant in this type of combined cycle can potentially be boosted to 50 percent or more.

Carbon dioxide is emitted as a concentrated gas stream in syngas at high pressure. In this form, it can be captured and sequestered more easily and at lower costs. By contrast, when coal burns or is reacted in air, 79 percent of which is nitrogen, the resulting carbon dioxide is diluted and more costly to separate.

Gasification Working

Gasification as Incomplete Combustion

Gasification is most simply thought of as choked combustion or incomplete combustion. It is burning solid fuels like wood or coal without enough air to complete combustion, so the output gas still has combustion potential. The unburned gas is then piped away to burn elsewhere as needed.

Gas produced by this method goes by a variety of names: wood gas, syngas, producer gas, town gas, generator gas, and others. It's sometimes also called biogas, though biogas more typically refers to gas produced via microbes in anaerobic digestion. In the context of biomass gasification using air-aspirated gasifiers, the term producer gas is the term we will be using, since the other terms have implications that do not necessarily apply to the gas produced by our gasifiers.

The Five Processes of Gasification

Gasification is more accurately understood as staged combustion. It is a series of distinct thermal events put together so as to purpose convert solid organic matter into specific hydrocarbon gases as output.

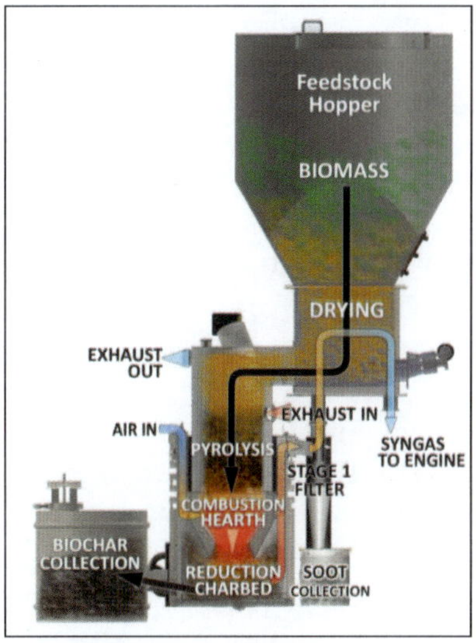

Simple incomplete combustion is a dirty mess. The goal in gasification is to take control of the discrete thermal processes usually mixed together in combustion, and reorganize them towards desired end products. In digital terms, "Gasification is the operating system of fire". Once you understand its underlying code, you can pull fire apart and reassemble it to your will, as well as a stunning variety of end products and processes.

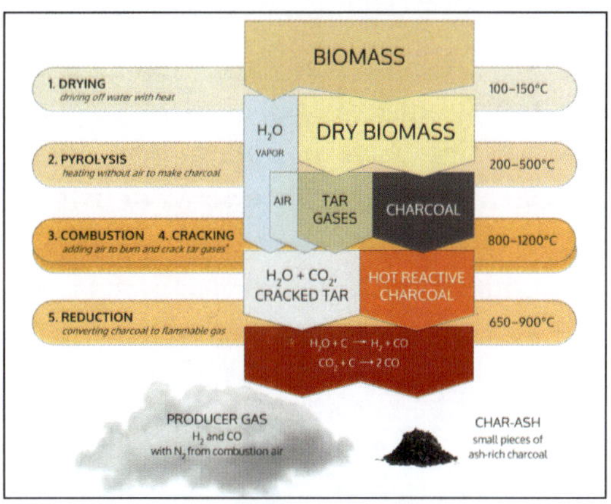

The five processes of gasification.

Gasification is made up for five discrete thermal processes: Drying, Pyrolysis, Combustion, Cracking, and Reduction. All of these processes are naturally present in the flame you see burning off a match, though they mix in a manner that renders them invisible to eyes not yet initiated into the mysteries of gasification. Gasification is merely the technology to pull apart and isolate these separate processes, so that we might interrupt the "fire" and pipe the resulting gases elsewhere.

Three of these processes tend to confuse all newcomers to gasification. Once you understand these three processes, all the others pieces fall in place quickly. These three non-obvious processes are Pyrolysis, Cracking, and Reduction.

Pyrolysis

Pyrolysis is the application of heat to raw biomass, in an absence of air, so as to break it down into charcoal and various tar gasses and liquids. It is essentially the process of charring.

Biomass begins to rapidly decompose with heat once its temperature rises above around 240°C. The biomass breaks down into a combination of solids, liquids and gasses. The solids that remain we commonly call charcoal. The gasses and liquids that are released we collectively call tars.

The gasses and liquids produced during lower temp pyrolysis are simply fragments of the original biomass that break off with heat. These fragments are the more complicated H, C and O molecules in the biomass that we collectively refer to as volatiles. As the name suggests, volatiles are reactive. Or more accurately, they are less strongly bonded in the biomass than the fixed carbon, which is the direct C to C bonds.

The input to gasification is some form of solid carbonaceous material— typically biomass or coal. All organic carbonaceous material is made up of carbon (C), hydrogen (H), an oxygen (O) atoms– though in a dizzying variety of molecular forms. The goal in gasification is to break down this wide variety of forms into the simple fuel gasses of H2 and CO— hydrogen and carbon monoxide.

Both hydrogen and carbon monoxide are burnable fuel gasses. We do not usually think of carbon monoxide as a fuel gas, but it actually has very good combustion characteristics (despite its poor characteristics when interacting with human hemoglobin). Carbon monoxide and hydrogen have about the same energy density by volume. Both are very clean burning as they only need to take on one oxygen atom, in one simple step, to arrive at the proper end states of combustion, CO_2 and H_2O. This is why an engine run on producer gas can have such clean emissions. The engine becomes the "afterburner" for the more dirty and difficult early stages of combustion that now are handled in the gasifier.

Thus, pyrolysis is the application of heat to biomass in the absence of air/oxygen. The volatiles in the biomass are evaporated off as tar gases, and the fixed carbon-to-carbon chains are what remains— otherwise known as charcoal.

Cracking

Cracking is the process of breaking down large complex molecules such as tar into lighter gases by exposure to heat. This process is crucial for the production of clean gas that is compatible with an internal combustion engine because tar gases condense into sticky tar that will rapidly foul the valves of an engine. Cracking is also necessary to ensure proper combustion because complete combustion only occurs when combustible gases thoroughly mix with oxygen. In the course of combustion, the high temperatures produced decompose the large tar molecules that pass through the combustion zone.

Reduction

Reduction is the process of stripping oxygen atoms off combustion products of hydrocarbon (HC) molecules, so as to return the molecules to forms that can burn again. Reduction is the direct reverse process of combustion. Combustion is the combination of combustible gases with oxygen to release heat, producing water vapor and carbon dioxide as waste products. Reduction is the removal of oxygen from these waste products at high temperature to produce combustable gases. Combustion and Reduction are equal and opposite reactions. In fact, in most burning environments, they are both operating simultaneously, in some form of dynamic equilibrium, with repeated movement back and forth between the two processes.

Reduction in a gasifier is accomplished by passing carbon dioxide (CO_2) or water vapor (H_2O) across a bed of red hot charcoal (C). The carbon in the hot charcoal is highly reactive with oxygen; it has such a high oxygen affinity that it strips the oxygen off water vapor and carbon dioxide, and redistributes it to as many single bond sites as possible. The oxygen is more attracted to the bond site on the C than to itself, thus no free oxygen can survive in its usual diatomic O_2 form. All available oxygen will bond to available C sites as individual O until all the oxygen is gone. When all the available oxygen is redistributed as single atoms, reduction stops.

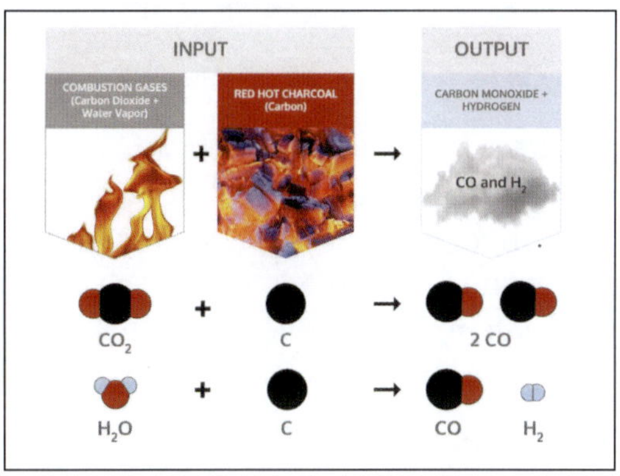

The reduction reactions.

Through this process, CO_2 is reduced by carbon to produce two CO molecules, and H_2O is reduced by carbon to produce H_2 and CO. Both H_2 and CO are combustable fuel gases, and those fuel gasses can then be piped off to do desired work elsewhere.

Combustion and Drying

These are the most easily understood of the Five Processes of Gasification. They do what we think by common understanding, though now they do it in the service of Pyrolysis and Reduction.

Combustion is the only net exothermic process of the Five Processes of Gasification; ultimately, all of the heat that drives drying, pyrolysis, and reduction comes either directly from combustion, or is recovered indirectly from combustion by heat exchange processes in a gasifier. Combustion can be fueled by either the tar gasses or char from Pyrolysis. Different reactor types use one or the other or both. In a downdraft gasifier, we are trying to burn the tar gasses from pyrolysis to generate heat to run reduction, as well as the CO_2 and H_2O to reduce in reduction. The goal in combustion in a downdraft is to get good mixing and high temps so that all the tars are either burned or cracked, and thus will not be present in the outgoing gas. The char bed and reduction contribute a relatively little to the conversion of messy tars to useful fuel gasses. Solving the tar problem is mostly an issue of tar cracking in the combustion zone.

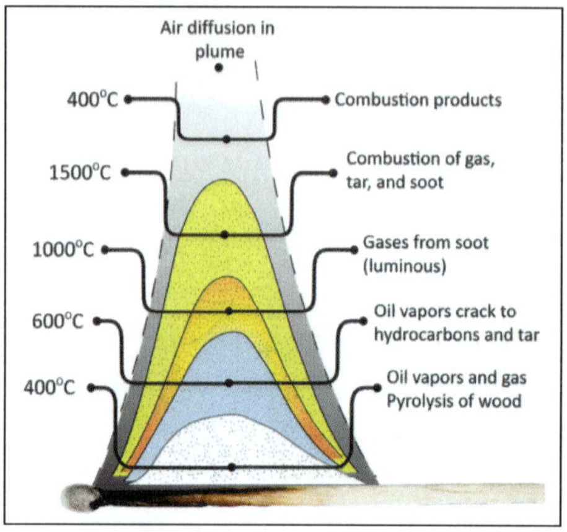

Pyrolysis, gasification and combustion in a flaming match.

Drying is what removes the moisture in the biomass before it enters Pyrolysis. All the moisture needs to be (or will be) removed from the fuel before any above 100°C processes happen. All of the water in the biomass will get vaporized out of the fuel at some point in the higher temp processes. Where and how this happens is one of the major issues that has to be solved for successful gasification. High moisture content fuel, and/or poor handling of the moisture internally, is one of the most common reasons for failure to produce clean gas.

More simply you might just think of gasification as burning a match, but interrupting the process by piping off the clear gas you see right above the match, not letting it mix with oxygen and complete combustion. Or you might think of it as running your car engine extremely rich, creating enough heat to break apart the raw fuel, but without enough oxygen to complete combustion, thus sending burnable gasses out the exhaust. This is how a hot rodder gets flames out the exhaust pipes.

CHEMICAL CONVERSION

Fischer–Tropsch Process

The Fischer–Tropsch process is a collection of chemical reactions that converts a mixture of carbon monoxide and hydrogen into liquid hydrocarbons. These reactions occur in the presence of metal catalysts, typically at temperatures of 150–300 °C (302–572 °F) and pressures of one to several tens of atmospheres. The process was first developed by Franz Fischer and Hans Tropsch at the Kaiser-Wilhelm-Institut für Kohlenforschung in Mülheim an der Ruhr, Germany, in 1925.

As a premier example of C1 chemistry, the Fischer–Tropsch process is an important reaction in both coal liquefaction and gas to liquids technology for producing liquid hydrocarbons. In the usual implementation, carbon monoxide and hydrogen, the feedstocks for FT, are produced from coal, natural gas, or biomass in a process known as gasification. The Fischer–Tropsch process then converts these gases into a synthetic lubrication oil and synthetic fuel. The Fischer–Tropsch process has received intermittent attention as a source of low-sulfur diesel fuel and to address the supply or cost of petroleum-derived hydrocarbons.

Reaction Mechanism

The Fischer–Tropsch process involves a series of chemical reactions that produce a variety of hydrocarbons, ideally having the formula (C_nH_{2n+2}). The more useful reactions produce alkanes as follows:

$$(2n + 1)\ H_2 + n\ CO \rightarrow C_nH_{2n+2} + n\ H_2O$$

where n is typically 10–20. The formation of methane ($n = 1$) is unwanted. Most of the alkanes produced tend to be straight-chain, suitable as diesel fuel. In addition to alkane formation, competing reactions give small amounts of alkenes, as well as alcohols and other oxygenated hydrocarbons.

Fischer–Tropsch Intermediates and Elemental Reactions

Converting a mixture of H_2 and CO into aliphatic products is a multi-step reaction with several intermediate compounds. The growth of the hydrocarbon chain may be visualized as involving a repeated sequence in which hydrogen atoms are added to carbon and oxygen, the C–O bond is split and a new C–C bond is formed. For one $-CH_2-$ group produced by $CO + 2\ H_2 \rightarrow (CH_2) + H_2O$, several reactions are necessary:

- Associative adsorption of CO.
- Splitting of the C–O bond.
- Dissociative adsorption of 2 H_2.
- Transfer of 2 H to the oxygen to yield H_2O.
- Desorption of H_2O.
- Transfer of 2 H to the carbon to yield CH_2.

The conversion of CO to alkanes involves hydrogenation of CO, the hydrogenolysis (cleavage with H_2) of C–O bonds, and the formation of C–C bonds. Such reactions are assumed to proceed via initial formation of surface-bound metal carbonyls. The CO ligand is speculated to undergo dissociation, possibly into oxide and carbide ligands. Other potential intermediates are various C_1 fragments including formyl (CHO), hydroxycarbene (HCOH), hydroxymethyl (CH_2OH), methyl (CH_3), methylene (CH_2), methylidyne (CH), and hydroxymethylidyne (COH). Furthermore, and critical to the production of liquid fuels, are reactions that form C–C bonds, such as migratory insertion. Many related stoichiometric reactions have been simulated on discrete metal clusters, but homogeneous Fischer–Tropsch catalysts are poorly developed and of no commercial importance.

Addition of isotopically labelled alcohol to the feed stream results in incorporation of alcohols into product. This observation establishes the facility of C–O bond scission. Using ^{14}C-labelled ethylene and propene over cobalt catalysts results in incorporation of these olefins into the growing chain. Chain growth reaction thus appears to involve both 'olefin insertion' as well as 'CO-insertion'.

Feedstocks: Gasification

Fischer–Tropsch plants associated with coal or related solid feedstocks (sources of carbon) must first convert the solid fuel into gaseous reactants, *i.e.*, CO, H_2, and alkanes. This conversion is called gasification and the product is called synthesis gas ("syngas"). Synthesis gas obtained from coal gasification tends to have a H_2:CO ratio of ~0.7 compared to the ideal ratio of ~2. This ratio is adjusted via the water-gas shift reaction. Coal-based Fischer–Tropsch plants produce varying amounts of CO_2, depending upon the energy source of the gasification process. However, most coal-based plants rely on the feed coal to supply all the energy requirements of the Fischer–Tropsch process.

Feedstocks: GTL

Carbon monoxide for FT catalysis is derived from hydrocarbons. In gas to liquids (GTL) technology, the hydrocarbons are low molecular weight materials that often would be discarded or flared. Stranded gas provides relatively cheap gas. GTL is viable provided gas remains relatively cheaper than oil.

Several reactions are required to obtain the gaseous reactants required for Fischer–Tropsch catalysis. First, reactant gases entering a Fischer–Tropsch reactor must be desulfurized. Otherwise, sulfur-containing impurities deactivate ("poison") the catalysts required for Fischer–Tropsch reactions.

Several reactions are employed to adjust the H_2:CO ratio. Most important is the water-gas shift reaction, which provides a source of hydrogen at the expense of carbon monoxide:

$$H_2O + CO \rightarrow H_2 + CO_2$$

For Fischer–Tropsch plants that use methane as the feedstock, another important reaction is steam reforming, which converts the methane into CO and H_2:

$$H_2O + CH_4 \rightarrow CO + 3\,H_2$$

Process Conditions

Generally, the Fischer–Tropsch process is operated in the temperature range of 150–300 °C (302–572 °F). Higher temperatures lead to faster reactions and higher conversion rates but also tend to favor methane production. For this reason, the temperature is usually maintained at the low to middle part of the range. Increasing the pressure leads to higher conversion rates and also favors formation of long-chained alkanes, both of which are desirable. Typical pressures range from one to several tens of atmospheres. Even higher pressures would be favorable, but the benefits may not justify the additional costs of high-pressure equipment, and higher pressures can lead to catalyst deactivation via coke formation.

A variety of synthesis-gas compositions can be used. For cobalt-based catalysts the optimal H_2:CO ratio is around 1.8–2.1. Iron-based catalysts can tolerate lower ratios, due to intrinsic water-gas shift reaction activity of the iron catalyst. This reactivity can be important for synthesis gas derived from coal or biomass, which tend to have relatively low H_2:CO ratios (< 1).

Design of the Fischer–Tropsch Process Reactor

Efficient removal of heat from the reactor is the basic need of Fischer–Tropsch reactors since these reactions are characterized by high exothermicity. Four types of reactors are discussed:

1. Multi tubular fixed-bed reactor:

This type of reactor contains a number of tubes with small diameter. These tubes contain catalyst and are surrounded by boiling water which removes the heat of reaction. A fixed-bed reactor is suitable for operation at low temperatures and has an upper temperature limit of 257 °C (530 K). Excess temperature leads to carbon deposition and hence blockage of the reactor. Since large amounts of the products formed are in liquid state, this type of reactor can also be referred to as a trickle flow reactor system.

2. Entrained flow Reactor:

An important requirement of the reactor for the Fischer–Tropsch process is to remove the heat of the reaction. This type of reactor contains two banks of heat exchangers which remove heat; the remainder of which is removed by the products and recycled in the system. The formation of heavy waxes should be avoided, since they condense on the catalyst and form agglomerations. This leads to fluidization. Hence, risers are operated over 297 °C (570 K).

3. Slurry Reactors:

Heat removal is done by internal cooling coils. The synthesis gas is bubbled through the waxy products and finely-divided catalyst which is suspended in the liquid medium. This also provides agitation of the contents of the reactor. The catalyst particle size reduces diffusional heat and mass transfer limitations. A lower temperature in the reactor leads to a more viscous product and a higher temperature (> 297 °C, 570 K) gives an undesirable product spectrum. Also, separation of the product from the catalyst is a problem.

4. Fluid-bed and Circulating Catalyst (Riser) Reactors:

These are used for high-temperature Fischer–Tropsch synthesis (nearly 340 °C) to produce low-molecular-weight unsaturated hydrocarbons on alkalized fused iron catalysts. The fluid-bed technology (as adapted from the catalytic cracking of heavy petroleum distillates) was introduced by Hydrocarbon Research in 1946–50 and named the 'Hydrocol' process. A large scale Fischer–Tropsch Hydrocol plant (350,000 tons per annum) operated during 1951–57 in Brownsville, Texas. Due to technical problems, and lacking economy due to increasing petroleum availability, this development was discontinued. Fluid-bed Fischer–Tropsch synthesis has recently been very successfully reinvestigated by Sasol. One reactor with a capacity of 500,000 tons per annum is now in operation and even larger ones are being built (nearly 850,000 tons per annum). The process is now used mainly for C_2 and C_7 alkene production.

This new development can be regarded as an important progress in Fischer–Tropsch technology. A high-temperature process with a circulating iron catalyst ('circulating fluid bed', 'riser reactor', 'entrained catalyst process') was introduced by the Kellogg Company and a respective plant built at Sasol in 1956. It was improved by Sasol for successful operation. At Secunda, South Africa, Sasol operated 16 advanced reactors of this type with a capacity of approximately 330,000 tons per annum each. Now the circulating catalyst process is being replaced by the superior Sasol-advanced fluid-bed technology. Early experiments with cobalt catalyst particles suspended in oil have been performed by Fischer. The bubble column reactor with a powdered iron slurry catalyst and a CO-rich syngas was particularly developed to pilot plant scale by Kölbel at the Rheinpreuben Company in 1953. Recently (since 1990) low-temperature Fischer–Tropsch slurry processes are under investigation for the use of iron and cobalt catalysts, particularly for the production of a hydrocarbon wax, or to be hydrocracked and isomerised to produce diesel fuel, by Exxon and Sasol. Today slurry-phase (bubble column) low-temperature Fischer–Tropsch synthesis is regarded by many authors as the most efficient process for Fischer–Tropsch clean diesel production. This Fischer–Tropsch technology is also under development by the Statoil Company (Norway) for use on a vessel to convert associated gas at offshore oil fields into a hydrocarbon liquid.

Product Distribution

In general the product distribution of hydrocarbons formed during the Fischer–Tropsch process follows an Anderson–Schulz–Flory distribution, which can be expressed as:

$$\frac{W_n}{n} = (1-\alpha)^2 \alpha^{n-1}$$

where W_n is the weight fraction of hydrocarbons containing n carbon atoms, and α is the chain growth probability or the probability that a molecule will continue reacting to form a longer chain. In general, α is largely determined by the catalyst and the specific process conditions.

Examination of the above equation reveals that methane will always be the largest single product so long as α is less than 0.5; however, by increasing α close to one, the total amount of methane formed can be minimized compared to the sum of all of the various long-chained products. Increasing α increases the formation of long-chained hydrocarbons. The very long-chained hydrocarbons are waxes, which are solid at room temperature. Therefore, for production of liquid transportation fuels it may be necessary to crack some of the Fischer–Tropsch products. In order to avoid this, some researchers have proposed using zeolites or other catalyst substrates with fixed sized pores that can restrict the formation of hydrocarbons longer than some characteristic size (usually $n < 10$). This way they can drive the reaction so as to minimize methane formation without producing lots of long-chained hydrocarbons. Such efforts have had only limited success.

Catalysts

A variety of catalysts can be used for the Fischer–Tropsch process, the most common are the transition metals cobalt, iron, and ruthenium. Nickel can also be used, but tends to favor methane formation ("methanation").

Cobalt

Cobalt-based catalysts are highly active, although iron may be more suitable for certain applications. Cobalt catalysts are more active for Fischer–Tropsch synthesis when the feedstock is natural gas. Natural gas has a high hydrogen to carbon ratio, so the water-gas shift is not needed for cobalt catalysts. Iron catalysts are preferred for lower quality feedstocks such as coal or biomass. Synthesis gases derived from these hydrogen-poor feedstocks has a low-hydrogen-content and require the water-gas shift reaction. Unlike the other metals used for this process (Co, Ni, Ru), which remain in the metallic state during synthesis, iron catalysts tend to form a number of phases, including various oxides and carbides during the reaction. Control of these phase transformations can be important in maintaining catalytic activity and preventing breakdown of the catalyst particles.

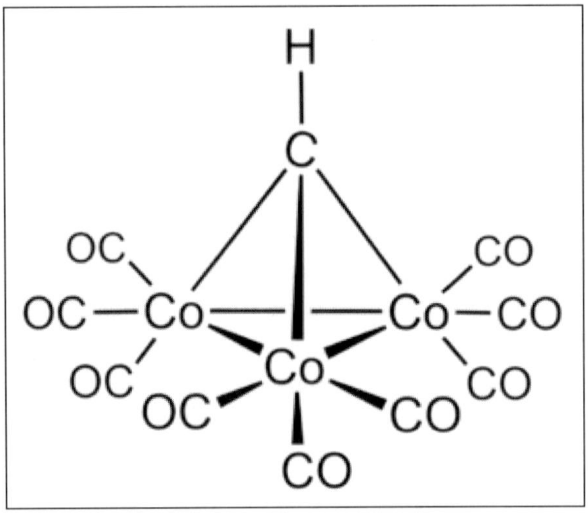

Methylidynetricobaltnonacarbonyl is a molecule that illustrates the kind of reduced carbon species speculated to occur in the Fischer–Tropsch process.

In addition to the active metal the catalysts typically contain a number of "promoters," including potassium and copper. Group 1 alkali metals, including potassium, are a poison for cobalt catalysts but are promoters for iron catalysts. Catalysts are supported on high-surface-area binders/supports such as silica, alumina, or zeolites. Promotors also have an important influence on activity. Alkali metal oxides and copper are common promotors, but the formulation depends on the primary metal, iron vs cobalt. Alkali oxides on cobalt catalysts generally cause activity to drop severely even with very low alkali loadings. $C \geq 5$ and CO_2 selectivity increase while methane and C_2–C_4 selectivity decrease. In addition, the alkene to alkane ratio increases.

Fischer–Tropsch catalysts are sensitive to poisoning by sulfur-containing compounds. Cobalt-based catalysts are more sensitive than their iron counterparts.

Iron

Fischer–Tropsch iron catalysts need alkali promotion to attain high activity and stability (e.g. 0.5 wt% K_2O). Addition of Cu for reduction promotion, addition of SiO_2, Al_2O_3 for structural promotion and maybe some manganese can be applied for selectivity control (e.g. high olefinicity). The working catalyst is only obtained when—after reduction with hydrogen—in the initial period of synthesis several iron carbide phases and elemental carbon are formed whereas iron oxides are still present in addition to some metallic iron. With iron catalysts two directions of selectivity have been pursued. One direction has aimed at a low-molecular-weight olefinic hydrocarbon mixture to be produced in an entrained phase or fluid bed process (Sasol–Synthol process). Due to the relatively high reaction temperature (approx. 340 °C), the average molecular weight of the product is so low that no liquid product phase occurs under reaction conditions. The catalyst particles moving around in the reactor are small (particle diameter 100 μm) and carbon deposition on the catalyst does not disturb reactor operation. Thus a low catalyst porosity with small pore diameters as obtained from fused magnetite (plus promoters) after reduction with hydrogen is appropriate. For maximising the overall gasoline yield, C_3 and C_4 alkenes have been oligomerized at Sasol. However, recovering the olefins for use as chemicals in, e.g., polymerization processes is advantageous today. The second direction of iron catalyst development has aimed at highest catalyst activity to be used at low reaction temperature where most of the hydrocarbon product is in the liquid phase under reaction conditions. Typically, such catalysts are obtained through precipitation from nitrate solutions. A high content of a carrier provides mechanical strength and wide pores for easy mass transfer of the reactants in the liquid product filling the pores. The main product fraction then is a paraffin wax, which is refined to marketable wax materials at Sasol; however, it also can be very selectively hydrocracked to a high quality diesel fuel. Thus, iron catalysts are very flexible.

Ruthenium

Ruthenium is the most active of the FT catalysts. It works at the lowest reaction temperatures, and it produces the highest molecular weight hydrocarbons. It acts as a Fischer–Tropsch catalyst as the pure metal, without any promotors, thus providing the simplest catalytic system of Fischer–Tropsch synthesis, where mechanistic conclusions should be the easiest—e.g., much easier than with iron as the catalyst. Like with nickel, the selectivity changes to mainly methane at elevated temperature. Its high price and limited world resources exclude industrial application. Systematic Fischer–Tropsch studies with ruthenium catalysts should contribute substantially to the further exploration of the fundamentals of Fischer–Tropsch synthesis. There is an interesting question to consider: what features have the metals nickel, iron, cobalt, and ruthenium in common to let them—and only them—be Fischer–Tropsch catalysts, converting the CO/H_2 mixture to aliphatic (long chain) hydrocarbons in a 'one step reaction'. The term 'one step reaction' means that reaction intermediates are not desorbed from the catalyst surface. In particular, it is amazing that the much carbided alkalized iron catalyst gives a similar reaction as the just metallic ruthenium catalyst.

HTFT and LTFT

High-Temperature Fischer–Tropsch (or HTFT) is operated at temperatures of 330–350 °C and uses an iron-based catalyst. This process was used extensively by Sasol in their coal-to-liquid plants (CTL). Low-Temperature Fischer–Tropsch (LTFT) is operated at lower temperatures and uses an iron or cobalt-based catalyst. This process is best known for being used in the first integrated GTL-plant operated and built by Shell in Bintulu, Malaysia.

Commercialization

Ras Laffan, Qatar

ORYX GTL Plant – Qatar.

The LTFT facility Pearl GTL at Ras Laffan, Qatar, is the largest FT plant. It uses cobalt catalysts at 230 °C, converting natural gas to petroleum liquids at a rate of 140,000 barrels per day (22,000 m^3/d), with additional production of 120,000 barrels (19,000 m^3) of oil equivalent in natural gas liquids and ethane. The plant in Ras Laffan was commissioned in 2007, called Oryx GTL, has a capacity of 34,000 barrels per day (5,400 m^3/d). The plant utilizes the Sasol slurry phase distillate process, which uses a cobalt catalyst. Oryx GTL is a joint venture between Qatar Petroleum and Sasol.

Sasol

A SASOL garage in Gauteng.

Another large scale implementation of Fischer–Tropsch technology is a series of plants operated by Sasol in South Africa, a country with large coal reserves, but little oil. The first commercial plant opened in 1952. Sasol uses coal and now natural gas as feedstocks and produces a variety of synthetic petroleum products, including most of the country's diesel fuel.

Sasol scrapped plans to build the GTL plant in Westlake, Louisiana.

PetroSA

PetroSA, another South African company, operates a refinery with a 36,000 barrels a day plant that completed semi-commercial demonstration in 2011, paving the way to begin commercial preparation. The technology can be used to convert natural gas, biomass or coal into synthetic fuels.

Shell Middle Distillate Synthesis

One of the largest implementations of Fischer–Tropsch technology is in Bintulu, Malaysia. This Shell facility converts natural gas into low-sulfur Diesel fuels and food-grade wax. The scale is 12,000 barrels per day (1,900 m^3/d).

Velocys

Construction is underway for Velocys' commercial reference plant incorporating its microchannel Fischer–Tropsch technology; ENVIA Energy's Oklahoma City GTL project being built adjacent to Waste Management's East Oak landfill site. The project is being financed by a joint venture between Waste Management, NRG Energy, Ventech and Velocys. The feedstock for this plant will be a combination of landfill gas and pipeline natural gas.

UPM (Finland)

In October 2006, Finnish paper and pulp manufacturer UPM announced its plans to produce biodiesel by the Fischer–Tropsch process alongside the manufacturing processes at its European paper and pulp plants, using waste biomass resulting from paper and pulp manufacturing processes as source material.

Rentech

A demonstration-scale Fischer–Tropsch plant was built and operated by Rentech, Inc., in partnership with ClearFuels, a company specializing in biomass gasification. Located in Commerce City, Colorado, the facility produces about 10 barrels per day (1.6 m^3/d) of fuels from natural gas. Commercial-scale facilities are planned for Rialto, California; Natchez, Mississippi; Port St. Joe, Florida; and White River, Ontario. Rentech closed down their pilot plant in 2013, and abandoned work on their FT process as well as the proposed commercial facilities.

Infra GTL Technology

In 2010, INFRA built a compact Pilot Plant for conversion of natural gas into synthetic oil. The plant modeled the full cycle of the GTL chemical process including the intake of pipeline gas, sulfur removal, steam methane reforming, syngas conditioning, and Fischer–Tropsch synthesis. In 2013 the first pilot plant was acquired by VNIIGAZ Gazprom LLC. In 2014 INFRA commissioned and operated on a continuous basis a new, larger scale full cycle Pilot Plant. It represents the second generation of INFRA's testing facility and is differentiated by a high degree of automation and extensive data gathering system. In 2015, INFRA built its own catalyst factory in Troitsk (Moscow, Russia). The catalyst factory has a capacity of over 15 tons per year, and produces the unique proprietary Fischer–Tropsch catalysts developed by the company's R&D division. In 2016, INFRA designed and built a modular, transportable GTL (gas-to-liquid) M100 plant for processing natural and associated gas into synthetic crude oil in Wharton (Texas, USA). The M100 plant is operating as a technology demonstration unit, R&D platform for catalyst refinement, and economic model to scale the Infra GTL process into larger and more efficient plants.

Other

In the United States and India, some coal-producing states have invested in Fischer–Tropsch plants. In Pennsylvania, Waste Management and Processors, Inc. was funded by the state to implement Fischer–Tropsch technology licensed from Shell and Sasol to convert so-called waste coal (leftovers from the mining process) into low-sulfur diesel fuel.

Process Efficiency

Using conventional FT technology the process ranges in carbon efficiency from 25 to 50 percent and a thermal efficiency of about 50% for CTL facilities idealised at 60% with GTL facilities at about 60% efficiency idealised to 80% efficiency.

Fischer–Tropsch in Nature

A Fischer–Tropsch-type process has also been suggested to have produced a few of the building blocks of DNA and RNA within asteroids. Similarly, the hypothetical abiogenic petroleum formation requires some naturally occurring FT-like processes.

BIOCHEMICAL CONVERSION OF BIOMASS

Biochemical conversion of biomass involves use of bacteria, microorganisms and enzymes to breakdown biomass into gaseous or liquid fuels, such as biogas or bioethanol.

The most popular biochemical technologies are anaerobic digestion (or biomethanation) and fermentation. Anaerobic digestion is a series of chemical reactions during which organic material is decomposed through the metabolic pathways of naturally occurring microorganisms in an oxygen depleted environment. Biomass wastes can also yield liquid fuels, such as cellulosic ethanol, which can be used to replace petroleum-based fuels.

Anaerobic digestion is the natural biological process which stabilizes organic waste in the absence of air and transforms it into biofertilizer and biogas. Anaerobic digestion is a reliable technology for the treatment of wet, organic waste. Organic waste from various sources is biochemically degraded in highly controlled, oxygen-free conditions circumstances resulting in the production of biogas which can be used to produce both electricity and heat. Almost any organic material can be processed with anaerobic digestion. This includes biodegradable waste materials such as municipal solid waste, animal manure, poultry litter, food wastes, sewage and industrial wastes.

An anaerobic digestion plant produces two outputs, biogas and digestate, both can be further processed or utilized to produce secondary outputs. Biogas can be used for producing electricity and heat, as a natural gas substitute and also a transportation fuel. A combined heat and power plant system (CHP) not only generates power but also produces heat for in-house requirements to maintain desired temperature level in the digester during cold season. In Sweden, the compressed biogas is used as a transportation fuel for cars and buses. Biogas can also be upgraded and used in gas supply networks.

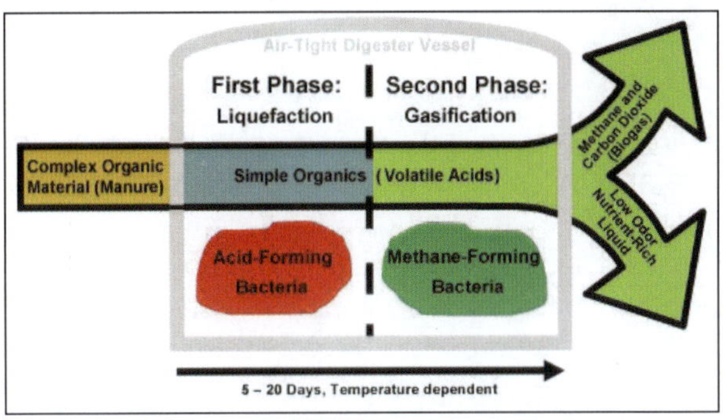

Working of Anaerobic Digestion Process.

Digestate can be further processed to produce liquor and a fibrous material. The fiber, which can be processed into compost, is a bulky material with low levels of nutrients and can be used as a soil conditioner or a low level fertilizer. A high proportion of the nutrients remain in the liquor, which can be used as a liquid fertilizer.

Biofuel Production

A variety of fuels can be produced from waste resources including liquid fuels, such as ethanol, methanol, biodiesel, Fischer-Tropsch diesel, and gaseous fuels, such as hydrogen and methane. The resource base for biofuel production is composed of a wide variety of forestry and agricultural resources, industrial processing residues, and municipal solid and urban wood residues. Globally, biofuels are most commonly used to power vehicles, heat homes, and for cooking, apart from powering home boilers.

The largest potential feedstock for ethanol is lignocellulosic biomass wastes, which includes materials such as agricultural residues (corn stover, crop straws and bagasse), herbaceous crops (alfalfa, switchgrass), short rotation woody crops, forestry residues, waste paper and other wastes (municipal and industrial). Bioethanol production from these feedstocks could be an attractive alternative for disposal of these residues. Importantly, lignocellulosic feedstocks do not interfere with food security.

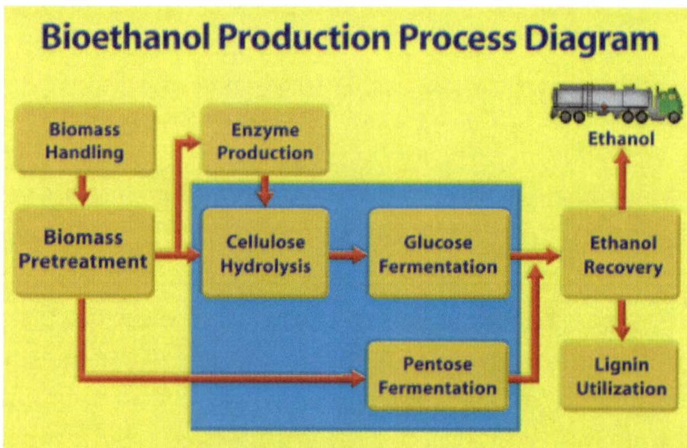

Ethanol from lignocellulosic biomass is produced mainly via biochemical routes. The three major steps involved are pretreatment, enzymatic hydrolysis, and fermentation. Biomass is pretreated to improve the accessibility of enzymes. After pretreatment, biomass undergoes enzymatic hydrolysis for conversion of polysaccharides into monomer sugars, such as glucose and xylose. Subsequently, sugars are fermented to ethanol by the use of different microorganisms.

Anaerobic Digestion

Anaerobic digestion is a collection of processes by which microorganisms break down biodegradable material in the absence of oxygen. The process is used for industrial or domestic purposes to manage waste or to produce fuels. Much of the fermentation used industrially to produce food and drink products, as well as home fermentation, uses anaerobic digestion.

Anaerobic digestion occurs naturally in some soils and in lake and oceanic basin sediments, where it is usually referred to as "anaerobic activity". This is the source of marsh gas methane as discovered by Alessandro Volta in 1776.

The digestion process begins with bacterial hydrolysis of the input materials. Insoluble organic polymers, such as carbohydrates, are broken down to soluble derivatives that become available for other bacteria. Acidogenic bacteria then convert the sugars and amino acids into carbon dioxide, hydrogen, ammonia, and organic acids. These bacteria convert these resulting organic acids into acetic acid, along with additional ammonia, hydrogen, and carbon dioxide. Finally, methanogens convert these products to methane and carbon dioxide. The methanogenic archaea populations play an indispensable role in anaerobic wastewater treatments.

Anaerobic digestion is used as part of the process to treat biodegradable waste and sewage sludge. As part of an integrated waste management system, anaerobic digestion reduces the emission of landfill gas into the atmosphere. Anaerobic digesters can also be fed with purpose-grown energy crops, such as maize.

Anaerobic digestion is widely used as a source of renewable energy. The process produces a biogas, consisting of methane, carbon dioxide, and traces of other 'contaminant' gases. This biogas can be used directly as fuel, in combined heat and power gas engines or upgraded to natural gas-quality biomethane. The nutrient-rich digestate also produced can be used as fertilizer.

With the re-use of waste as a resource and new technological approaches that have lowered capital costs, anaerobic digestion has in recent years received increased attention among governments in a number of countries, among these the United Kingdom, Germany and Denmark.

Process

Many microorganisms affect anaerobic digestion, including acetic acid-forming bacteria (acetogens) and methane-forming archaea (methanogens). These organisms promote a number of chemical processes in converting the biomass to biogas.

Gaseous oxygen is excluded from the reactions by physical containment. Anaerobes utilize electron acceptors from sources other than oxygen gas. These acceptors can be the organic material itself or may be supplied by inorganic oxides from within the input material. When the oxygen source in an anaerobic system is derived from the organic material itself, the 'intermediate' end products are primarily alcohols, aldehydes, and organic acids, plus carbon dioxide. In the presence of specialised methanogens, the intermediates are converted to the 'final' end products of methane, carbon dioxide, and trace levels of hydrogen sulfide. In an anaerobic system, the majority of the chemical energy contained within the starting material is released by methanogenic bacteria as methane.

Populations of anaerobic microorganisms typically take a significant period of time to establish themselves to be fully effective. Therefore, common practice is to introduce anaerobic microorganisms from materials with existing populations, a process known as "seeding" the digesters, typically accomplished with the addition of sewage sludge or cattle slurry.

Process Stages

The four key stages of anaerobic digestion involve hydrolysis, acidogenesis, acetogenesis and methanogenesis. The overall process can be described by the chemical reaction, where organic material such as glucose is biochemically digested into carbon dioxide (CO_2) and methane (CH_4) by the anaerobic microorganisms.

$$C_6H_{12}O_6 \rightarrow 3CO_2 + 3CH_4$$

Hydrolysis

In most cases, biomass is made up of large organic polymers. For the bacteria in anaerobic digesters to access the energy potential of the material, these chains must first be broken down into their smaller constituent parts. These constituent parts, or monomers, such as sugars, are readily available to other bacteria. The process of breaking these chains and dissolving the smaller molecules into solution is called hydrolysis. Therefore, hydrolysis of these high-molecular-weight polymeric components is the necessary first step in anaerobic digestion. Through hydrolysis the complex organic molecules are broken down into simple sugars, amino acids, and fatty acids.

Acetate and hydrogen produced in the first stages can be used directly by methanogens. Other molecules, such as volatile fatty acids (VFAs) with a chain length greater than that of acetate must first be catabolised into compounds that can be directly used by methanogens.

Acidogenesis

The biological process of acidogenesis results in further breakdown of the remaining components by acidogenic (fermentative) bacteria. Here, VFAs are created, along with ammonia, carbon dioxide, and hydrogen sulfide, as well as other byproducts. The process of acidogenesis is similar to the way milk sours.

Acetogenesis

The third stage of anaerobic digestion is acetogenesis. Here, simple molecules created through the acidogenesis phase are further digested by acetogens to produce largely acetic acid, as well as carbon dioxide and hydrogen.

Methanogenesis

The terminal stage of anaerobic digestion is the biological process of methanogenesis. Here, methanogens use the intermediate products of the preceding stages and convert them into methane, carbon dioxide, and water. These components make up the majority of the biogas emitted from the system. Methanogenesis is sensitive to both high and low pHs and occurs between pH 6.5 and pH 8. The remaining, indigestible material the microbes cannot use and any dead bacterial remains constitute the digestate.

Configuration

Anaerobic digesters can be designed and engineered to operate using a number of different configurations and can be categorized into batch vs. continuous process mode, mesophilic vs. thermophilic temperature conditions, high vs. low portion of solids, and single stage vs. multistage processes. Continuous process requires more complex design, but still, it may be more economical than batch process, because batch process requires more initial building money and a larger volume of the digesters (spread across several batches) to handle the same amount of waste as a continuous process digester. Higher heat energy is required in a thermophilic system compared to a mesophilic system, but the thermophilic system requires much less time and has a larger gas output capacity and higher methane gas content, so one has to consider that trade-off carefully. For solids content, low will handle up to 15% solid content. Above this level is considered high solids content and can also be known as dry digestion. In a single stage process, one reactor houses the four anaerobic digestion steps. A multistage process utilizes two or more reactors for digestion to separate the methanogenesis and hydrolysis phases.

Batch or Continuous

Anaerobic digestion can be performed as a batch process or a continuous process. In a batch system, biomass is added to the reactor at the start of the process. The reactor is then sealed for the duration of the process. In its simplest form batch processing needs inoculation with already processed material to start the anaerobic digestion. In a typical scenario, biogas production will be formed with a normal distribution pattern over time. Operators can use this fact to determine when they believe the process of digestion of the organic matter has completed. There can be severe odour issues if a batch reactor is opened and emptied before the process is well completed. A more advanced type of batch approach has limited the odour issues by integrating anaerobic digestion with in-vessel composting. In this approach inoculation takes place through the use of recirculated degasified percolate. After anaerobic digestion has completed, the biomass is kept in the reactor which is then used for in-vessel composting before it is opened As the batch digestion is simple and requires less equipment and lower levels of design work, it is typically a cheaper form of digestion. Using more than one batch reactor at a plant can ensure constant production of biogas.

In continuous digestion processes, organic matter is constantly added (continuous complete mixed) or added in stages to the reactor (continuous plug flow; first in – first out). Here, the end products are constantly or periodically removed, resulting in constant production of biogas. A single or multiple digesters in sequence may be used. Examples of this form of anaerobic digestion include continuous stirred-tank reactors, upflow anaerobic sludge blankets, expanded granular sludge beds, and internal circulation reactors.

Temperature

The two conventional operational temperature levels for anaerobic digesters determine the species of methanogens in the digesters:

- Mesophilic digestion takes place optimally around 30 to 38 °C, or at ambient temperatures between 20 and 45 °C, where mesophiles are the primary microorganism present.
- Thermophilic digestion takes place optimally around 49 to 57 °C, or at elevated temperatures up to 70 °C, where thermophiles are the primary microorganisms present.

A limit case has been reached in Bolivia, with anaerobic digestion in temperature working conditions of less than 10 °C. The anaerobic process is very slow, taking more than three times the normal mesophilic time process. In experimental work at University of Alaska Fairbanks, a 1,000 litre digester using psychrophiles harvested from "mud from a frozen lake in Alaska" has produced 200–300 litres of methane per day, about 20 to 30% of the output from digesters in warmer climates. Mesophilic species outnumber thermophiles, and they are also more tolerant to changes in environmental conditions than thermophiles. Mesophilic systems are, therefore, considered to be more stable than thermophilic digestion systems. In contrast, while thermophilic digestion systems are considered less stable, their energy input is higher, with more biogas being removed from the organic matter in an equal amount of time. The increased temperatures facilitate faster reaction rates, and thus faster gas yields. Operation at higher temperatures facilitates greater pathogen reduction of the digestate. In countries where legislation, such as the Animal By-Products Regulations in the European Union, requires digestate to meet certain levels of pathogen reduction there may be a benefit to using thermophilic temperatures instead of mesophilic.

Additional pre-treatment can be used to reduce the necessary retention time to produce biogas. For example, certain processes shred the substrates to increase the surface area or use a thermal pretreatment stage (such as pasteurisation) to significantly enhance the biogas output. The pasteurisation process can also be used to reduce the pathogenic concentration in the digesate leaving the anaerobic digester. Pasteurisation may be achieved by heat treatment combined with maceration of the solids.

Solids Content

In a typical scenario, three different operational parameters are associated with the solids content of the feedstock to the digesters:

- High solids (dry—stackable substrate)
- High solids (wet—pumpable substrate)
- Low solids (wet—pumpable substrate)

High solids (dry) digesters are designed to process materials with a solids content between 25 and 40%. Unlike wet digesters that process pumpable slurries, high solids (dry – stackable substrate) digesters are designed to process solid substrates without the addition of water. The primary styles of dry digesters are continuous vertical plug flow and batch tunnel horizontal digesters. Continuous vertical plug flow digesters are upright, cylindrical tanks where feedstock is continuously fed into the top of the digester, and flows downward by gravity during digestion. In batch tunnel digesters, the feedstock is deposited in tunnel-like chambers with a gas-tight door. Neither approach has mixing inside the digester. The amount of pretreatment, such as contaminant removal, depends both upon the nature of the waste streams being processed and the desired quality of the digestate. Size reduction (grinding) is beneficial in continuous vertical systems, as it accelerates digestion, while batch systems avoid grinding and instead require structure (e.g. yard waste) to reduce compaction of the stacked pile. Continuous vertical dry digesters have a smaller footprint due to the shorter effective retention time and vertical design. Wet digesters can be designed to operate in either a high-solids content, with a total suspended solids (TSS) concentration greater than ~20%, or a low-solids concentration less than ~15%.

High solids (wet) digesters process a thick slurry that requires more energy input to move and process the feedstock. The thickness of the material may also lead to associated problems with abrasion. High solids digesters will typically have a lower land requirement due to the lower volumes associated with the moisture. High solids digesters also require correction of conventional performance calculations (e.g. gas production, retention time, kinetics, etc.) originally based on very dilute sewage digestion concepts, since larger fractions of the feedstock mass are potentially convertible to biogas.

Low solids (wet) digesters can transport material through the system using standard pumps that require significantly lower energy input. Low solids digesters require a larger amount of land than high solids due to the increased volumes associated with the increased liquid-to-feedstock ratio of the digesters. There are benefits associated with operation in a liquid environment, as it enables more thorough circulation of materials and contact between the bacteria and their food. This enables the bacteria to more readily access the substances on which they are feeding, and increases the rate of gas production.

Complexity

Digestion systems can be configured with different levels of complexity. In a single-stage digestion system (one-stage), all of the biological reactions occur within a single, sealed reactor or holding tank. Using a single stage reduces construction costs, but results in less control of the reactions occurring within the system. Acidogenic bacteria, through the production of acids, reduce the pH of the tank. Methanogenic bacteria, as outlined earlier, operate in a strictly defined pH range. Therefore, the biological reactions of the different species in a single-stage reactor can be in direct competition with each other. Another one-stage reaction system is an anaerobic lagoon. These lagoons are pond-like, earthen basins used for the treatment and long-term storage of manures. Here the anaerobic reactions are contained within the natural anaerobic sludge contained in the pool.

In a two-stage digestion system (multistage), different digestion vessels are optimised to bring maximum control over the bacterial communities living within the digesters. Acidogenic bacteria produce organic acids and more quickly grow and reproduce than methanogenic bacteria. Methanogenic bacteria require stable pH and temperature to optimise their performance.

Under typical circumstances, hydrolysis, acetogenesis, and acidogenesis occur within the first reaction vessel. The organic material is then heated to the required operational temperature (either mesophilic or thermophilic) prior to being pumped into a methanogenic reactor. The initial hydrolysis or acidogenesis tanks prior to the methanogenic reactor can provide a buffer to the rate at which feedstock is added. Some European countries require a degree of elevated heat treatment to kill harmful bacteria in the input waste. In this instance, there may be a pasteurisation or sterilisation stage prior to digestion or between the two digestion tanks. Notably, it is not possible to completely isolate the different reaction phases, and often some biogas is produced in the hydrolysis or acidogenesis tanks.

Residence Time

The residence time in a digester varies with the amount and type of feed material, and with the configuration of the digestion system. In a typical two-stage mesophilic digestion, residence time varies between 15 and 40 days, while for a single-stage thermophilic digestion, residence times is normally faster and takes around 14 days. The plug-flow nature of some of these systems will mean the full degradation of the material may not have been realised in this timescale. In this event, digestate exiting the system will be darker in colour and will typically have more odour.

In the case of an upflow anaerobic sludge blanket digestion (UASB), hydraulic residence times can be as short as 1 hour to 1 day, and solid retention times can be up to 90 days. In this manner, a UASB system is able to separate solids and hydraulic retention times with the use of a sludge blanket. Continuous digesters have mechanical or hydraulic devices, depending on the level of solids in the material, to mix the contents, enabling the bacteria and the food to be in contact. They also allow excess material to be continuously extracted to maintain a reasonably constant volume within the digestion tanks.

Inhibition

The anaerobic digestion process can be inhibited by several compounds, affecting one or more of the bacterial groups responsible for the different organic matter degradation steps. The degree of the inhibition depends, among other factors, on the concentration of the inhibitor in the digester. Potential inhibitors are ammonia, sulfide, light metal ions (Na, K, Mg, Ca, Al), heavy metals, some organics (chlorophenols, halogenated aliphatics, N-substituted aromatics, long chain fatty acids), etc.

In figure, Left: Farm-based maize silage digester located near Neumünster in Germany, 2007 - the green, inflatable biogas holder is shown on top of the digester. Right: Two-stage, low solids, UASB digestion component of a mechanical biological treatment system near Tel Aviv; the process water is seen in balance tank and sequencing batch reactor, 2005.

Feedstocks

The most important initial issue when considering the application of anaerobic digestion systems is the feedstock to the process. Almost any organic material can be processed with anaerobic digestion; however, if biogas production is the aim, the level of putrescibility is the key factor in its successful application. The more putrescible (digestible) the material, the higher the gas yields possible from the system.

Anaerobic lagoon and generators at the Cal Poly Dairy, United States.

Feedstocks can include biodegradable waste materials, such as waste paper, grass clippings, leftover food, sewage, and animal waste. Woody wastes are the exception, because they are largely unaffected by digestion, as most anaerobes are unable to degrade lignin. Xylophalgeous anaerobes (lignin consumers) or using high temperature pretreatment, such as pyrolysis, can be used to break lignin down. Anaerobic digesters can also be fed with specially grown energy crops, such as silage, for dedicated biogas production. In Germany and continental Europe, these facilities are referred to as "biogas" plants. A codigestion or cofermentation plant is typically an agricultural anaerobic digester that accepts two or more input materials for simultaneous digestion.

The length of time required for anaerobic digestion depends on the chemical complexity of the material. Material rich in easily digestible sugars breaks down quickly, whereas intact lignocellulosic material rich in cellulose and hemicellulose polymers can take much longer to break down. Anaerobic microorganisms are generally unable to break down lignin, the recalcitrant aromatic component of biomass.

Anaerobic digesters were originally designed for operation using sewage sludge and manures. Sewage and manure are not, however, the material with the most potential for anaerobic digestion, as the biodegradable material has already had much of the energy content taken out by the animals that produced it. Therefore, many digesters operate with codigestion of two or more types of feedstock. For example, in a farm-based digester that uses dairy manure as the primary feedstock, the gas production may be significantly increased by adding a second feedstock, e.g., grass and corn (typical on-farm feedstock), or various organic byproducts, such as slaughterhouse waste, fats, oils and grease from restaurants, organic household waste, etc. (typical off-site feedstock).

Digesters processing dedicated energy crops can achieve high levels of degradation and biogas production. Slurry-only systems are generally cheaper, but generate far less energy than those using crops, such as maize and grass silage; by using a modest amount of crop material (30%), an anaerobic digestion plant can increase energy output tenfold for only three times the capital cost, relative to a slurry-only system.

Moisture Content

A second consideration related to the feedstock is moisture content. Drier, stackable substrates, such as food and yard waste, are suitable for digestion in tunnel-like chambers. Tunnel-style systems typically have near-zero wastewater discharge, as well, so this style of system has advantages where the discharge of digester liquids are a liability. The wetter the material, the more suitable it will be to handling with standard pumps instead of energy-intensive concrete pumps and physical means of movement. Also, the wetter the material, the more volume and area it takes up relative to the levels of gas produced. The moisture content of the target feedstock will also affect what type of system is applied to its treatment. To use a high-solids anaerobic digester for dilute feedstocks, bulking agents, such as compost, should be applied to increase the solids content of the input material. Another key consideration is the carbon:nitrogen ratio of the input material. This ratio is the balance of food a microbe requires to grow; the optimal C:N ratio is 20–30:1. Excess N can lead to ammonia inhibition of digestion.

Contamination

The level of contamination of the feedstock material is a key consideration. If the feedstock to the digesters has significant levels of physical contaminants, such as plastic, glass, or metals, then processing to remove the contaminants will be required for the material to be used. If it is not removed, then the digesters can be blocked and will not function efficiently. It is with this understanding that mechanical biological treatment plants are designed. The higher the level of pretreatment a feedstock requires, the more processing machinery will be required, and, hence, the project will have higher capital costs.

After sorting or screening to remove any physical contaminants from the feedstock, the material is often shredded, minced, and mechanically or hydraulically pulped to increase the surface area available to microbes in the digesters and, hence, increase the speed of digestion. The maceration of solids can be achieved by using a chopper pump to transfer the feedstock material into the airtight digester, where anaerobic treatment takes place.

Substrate Composition

Substrate composition is a major factor in determining the methane yield and methane production rates from the digestion of biomass. Techniques to determine the compositional characteristics of the feedstock are available, while parameters such as solids, elemental, and organic analyses are important for digester design and operation. Methane yield can be estimated from the elemental composition of substrate along with an estimate of its degradability (the fraction of the substrate that is converted to biogas in a reactor). In order to predict biogas composition (the relative fractions of methane and carbon dioxide) it is necessary to estimate carbon dioxide partitioning between the aqueous and gas phases, which requires additional information (reactor temperature, pH, and substrate composition) and a chemical speciation model. Direct measurements of biomethanation potential are also made using gas evolution or more recent gravimetric assays.

Applications

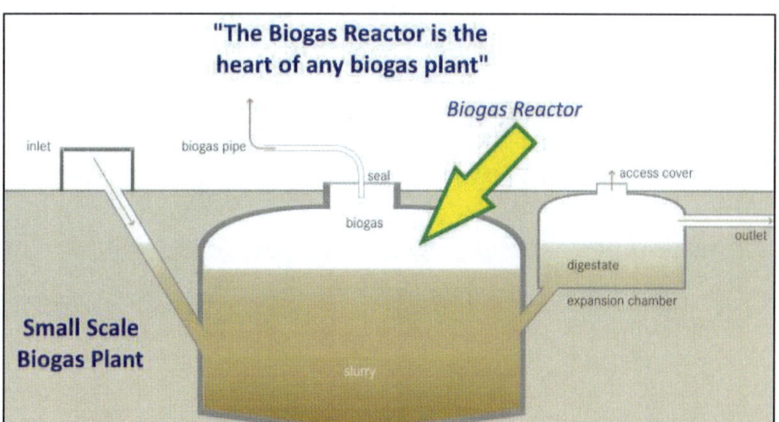

Schematic of an anaerobic digester as part of a sanitation system.

It produces a digested slurry (digestate) that can be used as a fertilizer, and biogas that can be used for energy.

Using anaerobic digestion technologies can help to reduce the emission of greenhouse gases in a number of key ways:

- Replacement of fossil fuels.
- Reducing or eliminating the energy footprint of waste treatment plants.
- Reducing methane emission from landfills.
- Displacing industrially produced chemical fertilizers.
- Reducing vehicle movements.
- Reducing electrical grid transportation losses.
- Reducing usage of LP Gas for cooking.
- An important component of the Zero Waste initiatives.

Waste and Wastewater Treatment

Anaerobic digesters in a sewage treatment plant. The methane gas is managed by burning through a gas flare.

Anaerobic digestion is particularly suited to organic material, and is commonly used for industrial effluent, wastewater and sewage sludge treatment. Anaerobic digestion, a simple process, can greatly reduce the amount of organic matter which might otherwise be destined to be dumped at sea, dumped in landfills, or burnt in incinerators.

Pressure from environmentally related legislation on solid waste disposal methods in developed countries has increased the application of anaerobic digestion as a process for reducing waste volumes and generating useful byproducts. It may either be used to process the source-separated fraction of municipal waste or alternatively combined with mechanical sorting systems, to process residual mixed municipal waste. These facilities are called mechanical biological treatment plants.

If the putrescible waste processed in anaerobic digesters were disposed of in a landfill, it would break down naturally and often anaerobically. In this case, the gas will eventually escape into the atmosphere. As methane is about 20 times more potent as a greenhouse gas than carbon dioxide, this has significant negative environmental effects.

In countries that collect household waste, the use of local anaerobic digestion facilities can help to reduce the amount of waste that requires transportation to centralized landfill sites or incineration facilities. This reduced burden on transportation reduces carbon emissions from the collection vehicles. If localized anaerobic digestion facilities are embedded within an electrical distribution network, they can help reduce the electrical losses associated with transporting electricity over a national grid.

Power Generation

In developing countries, simple home and farm-based anaerobic digestion systems offer the potential for low-cost energy for cooking and lighting. From 1975, China and India have both had large, government-backed schemes for adaptation of small biogas plants for use in the household for cooking and lighting. At present, projects for anaerobic digestion in the developing world can gain financial support through the United Nations Clean Development Mechanism if they are able to show they provide reduced carbon emissions.

Methane and power produced in anaerobic digestion facilities can be used to replace energy derived from fossil fuels, and hence reduce emissions of greenhouse gases, because the carbon in biodegradable material is part of a carbon cycle. The carbon released into the atmosphere from the combustion of biogas has been removed by plants for them to grow in the recent past, usually within the last decade, but more typically within the last growing season. If the plants are regrown, taking the carbon out of the atmosphere once more, the system will be carbon neutral. In contrast, carbon in fossil fuels has been sequestered in the earth for many millions of years, the combustion of which increases the overall levels of carbon dioxide in the atmosphere.

Biogas from sewage sludge treatment is sometimes used to run a gas engine to produce electrical power, some or all of which can be used to run the sewage works. Some waste heat from the engine is then used to heat the digester. The waste heat is, in general, enough to heat the digester to the required temperatures. The power potential from sewage works is limited – in the UK, there are about 80 MW total of such generation, with the potential to increase to 150 MW, which is insignificant compared to the average power demand in the UK of about 35,000 MW. The scope for biogas generation from nonsewage waste biological matter – energy crops, food waste, abattoir waste, etc. - is much higher, estimated to be capable of about 3,000 MW. Farm biogas plants using animal waste and energy crops are expected to contribute to reducing CO_2 emissions and strengthen the grid, while providing UK farmers with additional revenues.

Some countries offer incentives in the form of, for example, feed-in tariffs for feeding electricity onto the power grid to subsidize green energy production.

In Oakland, California at the East Bay Municipal Utility District's main wastewater treatment plant (EBMUD), food waste is currently codigested with primary and secondary municipal wastewater solids and other high-strength wastes. Compared to municipal wastewater solids digestion alone, food waste codigestion has many benefits. Anaerobic digestion of food waste pulp from the EBMUD food waste process provides a higher normalized energy benefit, compared to municipal wastewater solids: 730 to 1,300 kWh per dry ton of food waste applied compared to 560 to 940 kWh per dry ton of municipal wastewater solids applied.

Grid Injection

Biogas grid-injection is the injection of biogas into the natural gas grid. The raw biogas has to be previously upgraded to biomethane. This upgrading implies the removal of contaminants such as hydrogen sulphide or siloxanes, as well as the carbon dioxide. Several technologies are available for this purpose, the most widely implemented being pressure swing adsorption (PSA), water or amine scrubbing (absorption processes) and, in recent years, membrane separation. As an alternative, the electricity and the heat can be used for on-site generation, resulting in a reduction of losses in the transportation of energy. Typical energy losses in natural gas transmission systems range from 1–2%, whereas the current energy losses on a large electrical system range from 5–8%.

In October 2010, Didcot Sewage Works became the first in the UK to produce biomethane gas supplied to the national grid, for use in up to 200 homes in Oxfordshire. By 2017, UK electricity firm Ecotricity plan to have digester fed by locally sourced grass fueling 6000 homes.

Vehicle Fuel

After upgrading with the above-mentioned technologies, the biogas (transformed into biomethane) can be used as vehicle fuel in adapted vehicles. This use is very extensive in Sweden, where over 38,600 gas vehicles exist, and 60% of the vehicle gas is biomethane generated in anaerobic digestion plants.

Fertilizer and Soil Conditioner

The solid, fibrous component of the digested material can be used as a soil conditioner to increase the organic content of soils. Digester liquor can be used as a fertiliser to supply vital nutrients to soils instead of chemical fertilisers that require large amounts of energy to produce and transport. The use of manufactured fertilisers is, therefore, more carbon-intensive than the use of anaerobic digester liquor fertiliser. In countries such as Spain, where many soils are organically depleted, the markets for the digested solids can be equally as important as the biogas.

Cooking Gas

By using a bio-digester, which produces the bacteria required for decomposing, cooking gas is generated. The organic garbage like fallen leaves, kitchen waste, food waste etc. are fed into a crusher unit, where the mixture is conflated with a small amount of water. The mixture is then fed into the bio-digester, where the bacteria decomposes it to produce cooking gas. This gas is piped to kitchen stove. A 2 cubic meter bio-digester can produce 2 cubic meter of cooking gas. This is equivalent to 1 kg of LPG. The notable advantage of using a bio-digester is the sludge which is a rich organic manure.

Products

The three principal products of anaerobic digestion are biogas, digestate, and water.

Biogas

Typical composition of biogas		
Compound	Formula	%
Methane	CH_4	50–75
Carbon dioxide	CO_2	25–50
Nitrogen	N_2	0–10
Hydrogen	H_2	0–1
Hydrogen sulfide	H_2S	0–3
Oxygen	O_2	0–0

Biogas is the ultimate waste product of the bacteria feeding off the input biodegradable feedstock (the methanogenesis stage of anaerobic digestion is performed by archaea, a micro-organism on a distinctly different branch of the phylogenetic tree of life to bacteria), and is mostly methane and carbon dioxide, with a small amount hydrogen and trace hydrogen sulfide. (As-produced, biogas also contains water vapor, with the fractional water vapor volume a function of biogas temperature). Most of the biogas is produced during the middle of the digestion, after the bacterial population has grown, and tapers off as the putrescible material is exhausted. The gas is normally stored on top of the digester in an inflatable gas bubble or extracted and stored next to the facility in a gas holder.

The methane in biogas can be burned to produce both heat and electricity, usually with a reciprocating engine or microturbine often in a cogeneration arrangement where the electricity and waste heat generated are used to warm the digesters or to heat buildings. Excess electricity can be sold to suppliers or put into the local grid. Electricity produced by anaerobic digesters is considered to be renewable energy and may attract subsidies. Biogas does not contribute to increasing atmospheric carbon dioxide concentrations because the gas is not released directly into the atmosphere and the carbon dioxide comes from an organic source with a short carbon cycle.

Biogas may require treatment or 'scrubbing' to refine it for use as a fuel. Hydrogen sulfide, a toxic product formed from sulfates in the feedstock, is released as a trace component of the biogas. National environmental enforcement agencies, such as the U.S. Environmental Protection Agency or the English and Welsh Environment Agency, put strict limits on the levels of gases containing hydrogen sulfide, and, if the levels of hydrogen sulfide in the gas are high, gas scrubbing and cleaning equipment (such as amine gas treating) will be needed to process the biogas to within regionally accepted levels. Alternatively, the addition of ferrous chloride $FeCl_2$ to the digestion tanks inhibits hydrogen sulfide production.

Biogas holder with lightning protection rods and backup gas flare.

Volatile siloxanes can also contaminate the biogas; such compounds are frequently found in household waste and wastewater. In digestion facilities accepting these materials as a component of the feedstock, low-molecular-weight siloxanes volatilise into biogas. When this gas is combusted in a gas engine, turbine, or boiler, siloxanes are converted into silicon dioxide (SiO_2), which deposits internally in the machine, increasing wear and tear. Practical and cost-effective technologies to remove siloxanes and other biogas contaminants are available at the present time. In certain applications, *in situ* treatment can be used to increase the methane purity by reducing the offgas carbon dioxide content, purging the majority of it in a secondary reactor.

Biogas carrying pipes.

In countries such as Switzerland, Germany, and Sweden, the methane in the biogas may be compressed for it to be used as a vehicle transportation fuel or input directly into the gas mains. In countries where the driver for the use of anaerobic digestion are renewable electricity subsidies, this route of treatment is less likely, as energy is required in this processing stage and reduces the overall levels available to sell.

Digestate

Digestate is the solid remnants of the original input material to the digesters that the microbes cannot use. It also consists of the mineralised remains of the dead bacteria from within the digesters. Digestate can come in three forms: fibrous, liquor, or a sludge-based combination of the two fractions. In two-stage systems, different forms of digestate come from different digestion tanks. In single-stage digestion systems, the two fractions will be combined and, if desired, separated by further processing.

Acidogenic anaerobic digestate.

The second byproduct (acidogenic digestate) is a stable, organic material consisting largely of lignin and cellulose, but also of a variety of mineral components in a matrix of dead bacterial cells; some plastic may be present. The material resembles domestic compost and can be used as such or to make low-grade building products, such as fibreboard. The solid digestate can also be used as feedstock for ethanol production.

The third byproduct is a liquid (methanogenic digestate) rich in nutrients, which can be used as a fertiliser, depending on the quality of the material being digested. Levels of potentially toxic elements (PTEs) should be chemically assessed. This will depend upon the quality of the original feedstock. In the case of most clean and source-separated biodegradable waste streams, the levels of PTEs will be low. In the case of wastes originating from industry, the levels of PTEs may be higher and will need to be taken into consideration when determining a suitable end use for the material.

Digestate typically contains elements, such as lignin, that cannot be broken down by the anaerobic microorganisms. Also, the digestate may contain ammonia that is phytotoxic, and may hamper the growth of plants if it is used as a soil-improving material. For these two reasons, a maturation or composting stage may be employed after digestion. Lignin and other materials are available for degradation by aerobic microorganisms, such as fungi, helping reduce the overall volume of the material for transport. During this maturation, the ammonia will be oxidized into nitrates, improving the fertility of the material and making it more suitable as a soil improver. Large composting stages are typically used by dry anaerobic digestion technologies.

Wastewater

The final output from anaerobic digestion systems is water, which originates both from the moisture content of the original waste that was treated and water produced during the microbial reactions in the digestion systems. This water may be released from the dewatering of the digestate or may be implicitly separate from the digestate.

The wastewater exiting the anaerobic digestion facility will typically have elevated levels of biochemical oxygen demand (BOD) and chemical oxygen demand (COD). These measures of the reactivity of the effluent indicate an ability to pollute. Some of this material is termed 'hard COD', meaning it cannot be accessed by the anaerobic bacteria for conversion into biogas. If this effluent were put directly into watercourses, it would negatively affect them by causing eutrophication. As such, further treatment of the wastewater is often required. This treatment will typically be an oxidation stage wherein air is passed through the water in a sequencing batch reactors or reverse osmosis unit.

Fermentation

Fermentation is an anaerobic process that breaks down the glucose within organic materials. It is a series of chemical reactions that convert sugars to alcohol or acid. Yeast or bacteria are added to the biomass material, which feed on the sugars to produce ethanol and carbon dioxide. The ethanol is distilled and dehydrated to obtain a higher concentration of alcohol to achieve the required purity for the use as automotive fuel. The solid residue from the fermentation process can be used as cattle-feed and in the case of sugar cane; the bagasse can be used as a fuel for boilers or for subsequent gasification.

Fermentation of Forest and Industrial Residues

Raw materials containing sugars or be transformed into sugars can be used as fermentation substrates. The fermentable raw materials can be grouped as directly fermentable sugary materials, starchy, lignocellulosic materials and industrial wastes. Sugar containing materials require the least costly pretreatment, where starchy, lignocellulosic materials and urban wastes needed costly pretreatment, to convert into fermentable substrates.

Fermentation of Agricultural Wastes

Conventional crops such as corn and sugarcane are unable to meet the global demand of bioethanol production due to their primary value of food and feed. Therefore, lignocellulosic substances such as agricultural wastes are attractive feedstocks for bioethanol production. Agricultural wastes are cost effective, renewable and abundant. Bioethanol from agricultural waste could be a promising technology though the process has several challenges and limitations such as biomass transport and handling, and efficient pretreatment methods for total delignification of lignocellulosic. Proper pretreatment methods can increase concentrations of fermentable sugars after enzymatic saccharification, thereby improving the efficiency of the whole process. Conversion of glucose as well as xylose to ethanol needs some new fermentation technologies, to make the whole process cost effective. Goals of an effective pretreatment process are:

- Formation of sugars directly or subsequently by hydrolysis.
- To avoid loss and/or degradation of sugars formed.
- To limit formation of inhibitory products.
- To reduce energy demands.
- To minimize costs.

ELECTROCHEMICAL CONVERSION

Direct Carbon Fuel Cell

A Direct Carbon Fuel Cell (DCFC) is a fuel cell that uses a carbon rich material as a fuel such as bio-mass or coal. The cell produces energy by combining carbon and oxygen, which releases carbon dioxide as a by-product. It also called coal fuel cells (CFCs), carbon-air fuel cells (CAFCs), direct carbon/coal fuel cells (DCFCs), and DC-SOFC.

The total reaction of the cell is $C + O_2 \rightarrow CO_2$. The process in half cell notation:

- Anode: $C + 2\,O^{2-} \rightarrow CO_2 + 4\,e^-$
- Cathode: $O_2 + 4\,e^- \rightarrow 2\,O^{2-}$

Despite this release of carbon dioxide, the direct carbon fuel cell is more environmentally friendly than traditional carbon burning techniques. Due to its higher efficiency, it requires less carbon to produce the same amount of energy. Also, because pure carbon dioxide is emitted, carbon capture techniques are much cheaper than for conventional power stations. Utilized carbon can be in the form of coal, coke, char, or a non-fossilized source of carbon.

At least four types of DCFC exist:

- The first one is based on the Solid oxide fuel cell (SOFC) concept.

Overall reaction in the solid oxide electrolyte based DCFC,

$$C + O_2 \rightarrow CO_2$$

Anode reaction,

Direct electrochemical oxidation path,

$$C + 2O^{2-} \rightarrow CO_2 + 4e^-$$

$$C + O^{2-} \rightarrow CO + 2e^-$$

Indirect electrochemical oxidation path,

$$CO + O^{2-} \rightarrow CO_2 + 2e^-$$

Boudouard reaction: indirect chemical reaction path,

$$C + CO_2 \rightarrow 2CO$$

Cathode reaction,

$$O_2 + 4e^- \rightarrow 2O^{2-}$$

- The second one is molten hydroxides fuel cell. William W. Jacques obtained US Patent 555,511 in this type of fuel cell in 1896. Prototypes have been demonstrated by the research group, SARA, Inc.
- The third one is based on the Molten Carbonate Fuel Cell (MCFC) concept. William W. Jacques obtained a Canadian patent in this type of fuel cell in 1897. It has been developed further at the Lawrence Livermore Laboratory.
- The fourth is a molten tin anode solid oxide fuel cell design, which utilizes molten tin and tin oxide as an inter stage reaction between oxidation of the carbon dissolving in the anode and reduction of oxygen at the solid oxide cathode.

Direct Methanol Fuel Cell

Direct-methanol fuel cells or DMFCs are a subcategory of proton-exchange fuel cells in which methanol is used as the fuel. Their main advantage is the ease of transport of methanol, an energy-dense yet reasonably stable liquid at all environmental conditions.

Efficiency is quite low for these cells, reaching around 10%, so they are targeted especially to portable applications, where energy and power density are more important than efficiency.

A more efficient version of a direct fuel cell would play a key role in the theoretical use of methanol as a general energy transport medium, in the hypothesized methanol economy.

The Cell

In contrast to indirect methanol fuel cells, where methanol is reacted to hydrogen by steam reforming, DMFCs use a methanol solution (usually around 1M, i.e. about 3% in mass) to carry the reactant into the cell; common operating temperatures are in the range 50–120 °C, where high temperatures are usually pressurized. DMFCs themselves are more efficient at high temperatures and pressures, but these conditions end up causing so many losses in the complete system that the advantage is lost; therefore, atmospheric-pressure configurations are currently preferred.

Because of the methanol cross-over, a phenomenon by which methanol diffuses through the membrane without reacting, methanol is fed as a weak solution: this decreases efficiency significantly, since crossed-over methanol, after reaching the air side (the cathode), immediately reacts with air; though the exact kinetics are debated, the end result is a reduction of the cell voltage. Cross-over remains a major factor in inefficiencies, and often half of the methanol is lost to cross-over. Methanol cross-over and/or its effects can be alleviated by (a) developing alternative membranes (e.g.), (b) improving the electro-oxidation process in the catalyst layer and improving the structure of the catalyst and gas diffusion layers (e.g.), and (c) optimizing the design of the flow field and the membrane electrode assembly (MEA) which can be achieved by studying the current density distributions.

Other issues include the management of carbon dioxide created at the anode, the sluggish dynamic behavior, and the ability to maintain the solution water.

The only waste products with these types of fuel cells are carbon dioxide and water.

Application

Current DMFCs are limited in the power they can produce, but can still store a high energy content in a small space. This means they can produce a small amount of power over a long period of time. This makes them ill-suited for powering large vehicles (at least directly), but ideal for smaller vehicles such as forklifts and tuggers and consumer goods such as mobile phones, digital cameras or laptops. Military applications of DMFCs are an emerging application since they have low noise and thermal signatures and no toxic effluent. These applications include power for man-portable tactical equipment, battery chargers, and autonomous power for test and training instrumentation. Units are available with power outputs between 25 watts and 5 kilowatts with durations up to 100 hours between refuelings.

Methanol

Methanol is a liquid from -97.0 °C to 64.7 °C at atmospheric pressure. The energy density of methanol is an order of magnitude greater than even highly compressed hydrogen, and 15 times higher than Lithium-ion batteries.

Methanol is toxic and flammable. However, the International Civil Aviation Organization's (ICAO) Dangerous Goods Panel (DGP) voted in November 2005 to allow passengers to carry and use micro fuel cells and methanol fuel cartridges when aboard airplanes to power laptop computers and other consumer electronic devices. On September 24, 2007, the US Department of Transportation issued a proposal to allow airline passengers to carry fuel cell cartridges on board. The Department of Transportation issued a final ruling on April 30, 2008, permitting passengers and crew to carry an approved fuel cell with an installed methanol cartridge and up to two additional spare cartridges. It is worth noting that 200 ml maximum methanol cartridge volume allowed in the final ruling is double the 100 ml limit on liquids allowed by the Transportation Security Administration in carry-on bags.

Reaction

The DMFC relies upon the oxidation of methanol on a catalyst layer to form carbon dioxide. Water is consumed at the anode and is produced at the cathode. Protons (H^+) are transported across the proton exchange membrane - often made from Nafion - to the cathode where they react with oxygen to produce water. Electrons are transported through an external circuit from anode to cathode, providing power to connected devices.

The half-reactions are:

	Equation
Anode	$CH_3OH + H_2O \rightarrow 6\,H^+ + 6\,e^- + CO_2$ oxidation
Cathode	$\frac{3}{2}O_2 + 6\,H^+ + 6\,e^- \rightarrow 3\,H_2O$ reduction
Overall reaction	$CH_3OH + \frac{3}{2}O_2 \rightarrow 2\,H_2O + CO_2$ redox reaction

Methanol and water are adsorbed on a catalyst usually made of platinum and ruthenium particles, and lose protons until carbon dioxide is formed. As water is consumed at the anode in the reaction, pure methanol cannot be used without provision of water via either passive transport such as back diffusion (osmosis), or active transport such as pumping. The need for water limits the energy density of the fuel.

Platinum is used as a catalyst for both half-reactions. This contributes to the loss of cell voltage potential, as any methanol that is present in the cathode chamber will oxidize. If another catalyst could be found for the reduction of oxygen, the problem of methanol crossover would likely be significantly lessened. Furthermore, platinum is very expensive and contributes to the high cost per kilowatt of these cells.

During the methanol oxidation reaction carbon monoxide (CO) is formed, which strongly adsorbs onto the platinum catalyst, reducing the number of available reaction sites and thus the performance of the cell. The addition of other metals, such as ruthenium or gold, to the platinum catalyst tends to ameliorate this problem. In the case of platinum-ruthenium catalysts, the oxophilic nature of ruthenium is believed to promote the formation of hydroxyl radicals on its surface, which can then react with carbon monoxide adsorbed on the platinum atoms. The water in the fuel cell is oxidized to a hydroxy radical via the following reaction: $H_2O \rightarrow OH\bullet + H^+ + e^-$. The hydroxy radical then oxidizes carbon monoxide to produce carbon dioxide, which is released from the surface as a gas: $CO + OH\bullet \rightarrow CO_2 + H^+ + e^-$.

Using these OH groups in the half reactions, they are also expressed as:

	Equation
Anode	$CH_3OH + 6\,OH^- \rightarrow 5\,H_2O + 6\,e^- + CO_2$ oxidation
Cathode	$\frac{3}{2}O_2 + 3\,H_2O + 6\,e^- \rightarrow 6\,OH^-$ reduction
Overall reaction	$CH_3OH + \frac{3}{2}O_2 \rightarrow 2\,H_2O + CO_2$ redox reaction

Cross-over Current

Methanol on the anodic side is usually in a weak solution (from 1M to 3M), because methanol in high concentrations has the tendency to diffuse through the membrane to the cathode, where its concentration is about zero because it is rapidly consumed by oxygen. Low concentrations help in reducing the cross-over, but also limit the maximum attainable current.

The practical realization is usually that a solution loop enters the anode, exits, is refilled with methanol, and returns to the anode again. Alternatively, fuel cells with optimized structures can be directly fed with high concentration methanol solutions or even pure methanol.

Water Drag

The water in the anodic loop is lost because of the anodic reaction, but mostly because of the associated water drag: every proton formed at the anode drags a number of water molecules to the cathode. Depending on temperature and membrane type, this number can be between 2 and 6.

Ancillary Units

A direct methanol fuel cell is usually part of a larger system including all the ancillary units that permit its operation. Compared to most other types of fuel cells, the ancillary system of DMFCs is relatively complex. The main reasons for its complexity are:

- Providing water along with methanol would make the fuel supply more cumbersome, so water has to be recycled in a loop.
- CO_2 has to be removed from the solution flow exiting the fuel cell.
- Water in the anodic loop is slowly consumed by reaction and drag; it is necessary to recover water from the cathodic side to maintain steady operation.

Direct-ethanol Fuel Cell

Direct-ethanol fuel cells or DEFCs are a category of fuel cell in which ethanol is fed directly into the cell. They have been used as a model to investigate a range of fuel cell concepts including the use of PEM.

Advantages

DEFC uses Ethanol in the fuel cell instead of the more toxic methanol. Ethanol is an attractive alternative to methanol because it comes with a supply chain that's already in place. Ethanol also remains the easier fuel to work with for widespread use by consumers.

Ethanol is a hydrogen-rich liquid and it has a higher specific energy (8.0 kWh/kg) compared to methanol (6.1 kWh/kg). Ethanol can be obtained in great quantity from biomass through a fermentation process from renewable resources like from sugar cane, wheat, corn, or even straw. Bio-generated ethanol (or bio-ethanol) is thus attractive since growing crops for biofuels absorbs much of the carbon dioxide emitted into the atmosphere from fuel used to produce the biofuels, and from burning the biofuels themselves. This is in sharp contrast to the use of fossil fuels. The use of ethanol would also overcome both the storage and infrastructure challenge of hydrogen for fuel cell applications. In a fuel cell, the oxidation of any fuel requires the use of a catalyst in order to achieve the current densities required for commercially viable fuel cells, and platinum-based catalysts are some of the most efficient materials for the oxidation of small organic molecules.

Reaction

The DEFC, similar to the DMFC, relies upon the oxidation of ethanol on a catalyst layer to form carbon dioxide. Water is consumed at the anode and is produced at the cathode. Protons (H^+) are transported across the proton exchange membrane to the cathode where they react with oxygen to produce water. Electrons are transported through an external circuit from anode to cathode, providing power to connected devices.

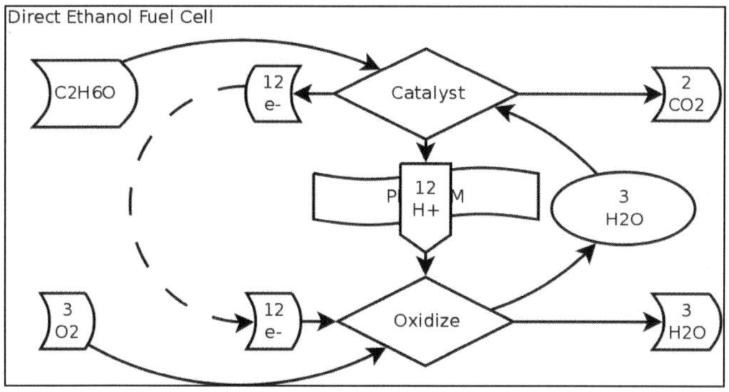

Flowchart of the reaction in a DEFC.

The half-reactions are:

	Equation
Anode	$C_2H_5OH + 3\,H_2O \rightarrow 12\,H^+ + 12\,e^- + 2\,CO_2$ oxidation
Cathode	$3\,O_2 + 12\,H^+ + 12\,e^- \rightarrow 6\,H_2O$ reduction
Overall reaction	$C_2H_5OH + 3\,O_2 \rightarrow 3\,H_2O + 2\,CO_2$ redox reaction

Issues

Platinum-based catalysts are expensive, so practical exploitation of ethanol as fuel for a PEM fuel cell requires a new catalyst. New nanostructured electrocatalysts (HYPERMEC by ACTA SpA for example) have been developed, which are based on non-noble metals, preferentially mixtures of Fe, Co, Ni at the anode, and Ni, Fe or Co alone at the cathode. With ethanol, power densities as high as 140 mW/cm² at 0.5 V have been obtained at 25 °C with self-breathing cells containing commercial anion exchange membranes. This catalyst does not contain any precious metals. In practice tiny metal particles are fixed onto a substrate in such a way that they produce a very active catalyst.

A polymer acts as electrolyte. The charge is carried by the hydrogen ion (proton). The liquid ethanol (C_2H_5OH) is oxidized at the anode in the presence of water, generating CO_2, hydrogen ions and electrons. Hydrogen ions travel through the electrolyte. They react at the cathode with oxygen from the air and the electrons from the external circuit forming water.

Bio-Ethanol based fuel cells may improve the well-to-wheel balance of this biofuel because of the increased conversion rate of the fuel cell compared to the internal combustion engine. But real world figures may be only achieved in some years since the development of direct methanol and ethanol fuel cells is lagging behind hydrogen powered fuel cells.

BIOCONVERSION OF BIOMASS TO MIXED ALCOHOL FUELS

The bioconversion of biomass to mixed alcohol fuels can be accomplished using the MixAlco process. Through bioconversion of biomass to a mixed alcohol fuel, more energy from the biomass will end up as liquid fuels than in converting biomass to ethanol by yeast fermentation.

The process involves a biological/chemical method for converting any biodegradable material (e.g., urban wastes, such as municipal solid waste, biodegradable waste, and sewage sludge, agricultural residues such as corn stover, sugarcane bagasse, cotton gin trash, manure) into useful chemicals, such as carboxylic acids (e.g., acetic, propionic, butyric acid), ketones (e.g., acetone, methyl ethyl ketone, diethyl ketone) and biofuels, such as a mixture of primary alcohols (e.g., ethanol, propanol, *n*-butanol) and/or a mixture of secondary alcohols (e.g., isopropanol, 2-butanol, 3-pentanol). Because of the many products that can be economically produced, this process is a true biorefinery.

Pilot Plant.

The process uses a mixed culture of naturally occurring microorganisms found in natural habitats such as the rumen of cattle, termite guts, and marine and terrestrial swamps to anaerobically digest biomass into a mixture of carboxylic acids produced during the acidogenic and acetogenic stages of anaerobic digestion, however with the inhibition of the methanogenic final stage. The more popular methods for production of ethanol and cellulosic ethanol use enzymes that must be isolated first to be added to the biomass and thus convert the starch or cellulose into simple sugars, followed then by yeast fermentation into ethanol. This process does not need the addition of such enzymes as these microorganisms make their own.

As the microoganisms anaerobically digest the biomass and convert it into a mixture of carboxylic acids, the pH must be controlled. This is done by the addition of a buffering agent (e.g., ammonium bicarbonate, calcium carbonate), thus yielding a mixture of carboxylate salts. Methanogenesis, being the natural final stage of anaerobic digestion, is inhibited by the presence of the ammonium ions or by the addition of an inhibitor (e.g., iodoform). The resulting fermentation broth contains the produced carboxylate salts that must be dewatered. This is achieved efficiently by vapor-compression evaporation. Further chemical refining of the dewatered fermentation broth may then take place depending on the final chemical or biofuel product desired.

The condensed distilled water from the vapor-compression evaporation system is recycled back to the fermentation. On the other hand, if raw sewage or other waste water with high BOD in need of treatment is used as the water for the fermentation, the condensed distilled water from the evaporation can be recycled back to the city or to the original source of the high-BOD waste water. Thus, this process can also serve as a water treatment facility, while producing valuable chemicals or biofuels.

Because the system uses a mixed culture of microorganisms, besides not needing any enzyme addition, the fermentation requires no sterility or aseptic conditions, making this front step in the process more economical than in more popular methods for the production of cellulosic ethanol. These savings in the front end of the process, where volumes are large, allows flexibility for further chemical transformations after dewatering, where volumes are small.

Carboxylic Acids

Carboxylic acids can be regenerated from the carboxylate salts using a process known as "acid springing". This process makes use of a high-molecular-weight tertiary amine (e.g., trioctylamine), which is switched with the cation (e.g., ammonium or calcium). The resulting amine carboxylate can then be thermally decomposed into the amine itself, which is recycled, and the corresponding carboxylic acid. In this way, theoretically, no chemicals are consumed or wastes produced during this step.

Ketones

There are two methods for making ketones. The first one consists on thermally converting calcium carboxylate salts into the corresponding ketones. This was a common method for making acetone from calcium acetate during World War I. The other method for making ketones consists on converting the vaporized carboxylic acids on a catalytic bed of zirconium oxide.

Alcohols

Primary Alcohols

The undigested residue from the fermentation may be used in gasification to make hydrogen (H_2). This H_2 can then be used to hydrogenolyze the esters over a catalyst (e.g., copper chromite), which are produced by esterifying either the ammonium carboxylate salts (e.g., ammonium acetate, propionate, butyrate) or the carboxylic acids (e.g., acetic, propionic, butyric acid) with a high-molecular-weight alcohol (e.g., hexanol, heptanol). From the hydrogenolysis, the final products are the high-molecular-weight alcohol, which is recycled back to the esterification, and the corresponding primary alcohols (e.g., ethanol, propanol, butanol).

Secondary Alcohols

The secondary alcohols (e.g., isopropanol, 2-butanol, 3-pentanol) are obtained by hydrogenating over a catalyst (e.g., Raney nickel) the corresponding ketones (e.g., acetone, methyl ethyl ketone, diethyl ketone).

Drop-in Biofuels

The primary or secondary alcohols obtained as described above may undergo conversion to drop-in biofuels, fuels which are compatible with current fossil fuel infrastructure such as biogasoline, green diesel and bio-jet fuel. Such is done by subjecting the alcohols to dehydration followed by oligomerization using zeolite catalysts in a manner similar to the methanex process, which used to produce gasoline from methanol in New Zealand.

Acetic Acid versus Ethanol

Cellulosic-ethanol manufacturing plants are bound to be net exporters of electricity because a large portion of the lignocellulosic biomass, namely lignin, remains undigested and it must be burned, thus producing electricity for the plant and excess electricity for the grid. As the market grows and this technology becomes more widespread, coupling the liquid fuel and the electricity markets will become more and more difficult.

Acetic acid, unlike ethanol, is biologically produced from simple sugars without the production of carbon dioxide:

$$C_6H_{12}O_6 \rightarrow 2\ CH_3CH_2OH + 2\ CO_2$$

(Biological production of ethanol)

$$C_6H_{12}O_6 \rightarrow 3\ CH_3COOH$$

(Biological production of acetic acid)

Because of this, on a mass basis, the yields will be higher than in ethanol fermentation. If then, the undigested residue (mostly lignin) is used to produce hydrogen by gasification, it is ensured that more energy from the biomass will end up as liquid fuels rather than excess heat/electricity.

$$3\ CH_3COOH + 6\ H_2 \rightarrow 3\ CH_3CH_2OH + 3\ H_2O$$

(Hydrogenation of acetic acid)

$$C_6H_{12}O_6\ \text{(from cellulose)} + 6\ H_2\ \text{(from lignin)} \rightarrow 3\ CH_3CH_2OH + 3\ H_2O$$

(Overall reaction)

A more comprehensive description of the economics of each of the fuels is given on the pages alcohol fuel and ethanol fuel.

Stage of Development

The system has been in development since 1991, moving from the laboratory scale (10 g/day) to the pilot scale (200 lb/day) in 2001. A small demonstration-scale plant (5 ton/day) has been constructed and is under operation and a 220 ton/day demonstration plant is expected in 2012.

THERMAL DEPOLYMERIZATION

Thermal depolymerization (TDP) is a depolymerization process using hydrous pyrolysis for the reduction of complex organic materials (usually waste products of various sorts, often biomass and plastic) into light crude oil. It mimics the natural geological processes thought to be involved in the production of fossil fuels. Under pressure and heat, long chain polymers of hydrogen, oxygen, and carbon decompose into short-chain petroleum hydrocarbons with a maximum length of around 18 carbons.

Similar Processes

Thermal depolymerisation is similar to other processes which use superheated water as a major step to produce fuels, such as direct Hydrothermal Liquefaction. These are distinct from processes using dry materials to depolymerize, such as pyrolysis. The term Thermochemical Conversion (TCC) has also been used for conversion of biomass to oils, using superheated water, although it is more usually applied to fuel production via pyrolysis. Other commercial scale processes include the "SlurryCarb" process operated by EnerTech, which uses similar technology to decarboxylate wet solid biowaste, which can then be physically dewatered and used as a solid fuel called E-Fuel. The plant in Rialto, California, was designed to process 683 tons of waste per day. However, it failed to perform to design standards and was closed down. The Rialto facility defaulted on its bond payments and is in the process of being liquidated. The Hydro Thermal Upgrading (HTU) process uses superheated water to produce oil from domestic waste. A demonstration plant is due to start up in The Netherlands said to be capable of processing

64 tons of biomass (dry basis) per day into oil. Thermal depolymerisation differs in that it contains a hydrous process followed by an anhydrous cracking / distillation process.

Theory and Process

In the method used by CWT, the water improves the heating process and contributes hydrogen to the reactions.

In the Changing World Technologies (CWT) process, the feedstock material is first ground into small chunks, and mixed with water if it is especially dry. It is then fed into a pressure vessel reaction chamber where it is heated at constant volume to around 250 °C. Similar to a pressure cooker (except at much higher pressure), steam naturally raises the pressure to 600 psi (4 MPa) (near the point of saturated water). These conditions are held for approximately 15 minutes to fully heat the mixture, after which the pressure is rapidly released to boil off most of the water.The result is a mix of crude hydrocarbons and solid minerals. The minerals are removed, and the hydrocarbons are sent to a second-stage reactor where they are heated to 500 °C, further breaking down the longer hydrocarbon chains. The hydrocarbons are then sorted by fractional distillation, in a process similar to conventional oil refining.

The CWT company claims that 15 to 20% of feedstock energy is used to provide energy for the plant. The remaining energy is available in the converted product. Working with turkey offal as the feedstock, the process proved to have yield efficiencies of approximately 85%; in other words, the energy contained in the end products of the process is 85% of the energy contained in the inputs to the process (most notably the energy content of the feedstock, but also including electricity for pumps and natural gas or woodgas for heating). If one considers the energy content of the feedstock to be free (i.e., waste material from some other process), then 85 units of energy are made available for every 15 units of energy consumed in process heat and electricity. This means the "Energy Returned on Energy Invested" (EROEI) is (6.67), which is comparable to other energy harvesting processes. Higher efficiencies may be possible with drier and more carbon-rich feedstocks, such as waste plastic.

By comparison, the current processes used to produce ethanol and biodiesel from agricultural sources have EROEI in the 4.2 range, when the energy used to produce the feedstocks is accounted for (in this case, usually sugar cane, corn, soybeans and the like). These EROEI values are not directly comparable, because these EROEI calculations include the energy cost to produce the feedstock, whereas the above EROEI calculation for thermal depolymerization process (TDP) does not.

The process breaks down almost all materials that are fed into it. TDP even efficiently breaks down many types of hazardous materials, such as poisons and difficult-to-destroy biological agents such as prions.

Feedstocks and Outputs with Thermal Depolymerization

Average TDP Feedstock Outputs				
Feedstock	Oils	Gas-es	Solids (mostly carbon based)	Water (Steam)
Plastic bottles	70%	16%	6%	8%
Medical waste	65%	10%	5%	20%
Tires	44%	10%	42%	4%
Turkey offal	39%	6%	5%	50%
Sewage sludge	26%	9%	8%	57%
Paper (cellulose)	8%	48%	24%	20%

Carthage Plant Products

As reported on 04/02/2006 by Discover Magazine, a Carthage, Missouri plant was producing 500 barrels per day (79 m^3/d) of oil made from 270 tons of turkey entrails and 20 tons of hog lard. This represents an oil yield of 22.3 percent. The Carthage plant produces API 40+, a high value crude oil. It contains light and heavy naphthas, a kerosene, and a gas oil fraction, with essentially no heavy fuel oils, tars, asphaltenes, or waxes. It can be further refined to produce No. 2 and No. 4 fuel oils.

TDP-40 Oil Classification by D-5443 PONA method	
Output Material	% by Weight
Paraffins	22%
Olefins	14%
Naphthenes	3%
Aromatics	6%
C14/C14+	55%
	100%

The fixed carbon solids produced by the TDP process have multiple uses as a filter, a fuel source, and a fertilizer. It can be used as activated carbon in wastewater treatment, as a fertilizer, or as a fuel similar to coal.

Advantages

The process can break down organic poisons, due to breaking chemical bonds and destroying the molecular shape needed for the poison's activity. It is likely to be highly effective at killing pathogens, including prions. It can also safely remove heavy metals from the samples by converting them from their ionized or organometallic forms to their stable oxides which can be safely separated from the other products.

Along with similar processes, it is a method of recycling the energy content of organic materials without first removing the water. It can produce liquid fuel, which separates from the water physically without need for drying. Other methods to recover energy often require pre-drying (e.g. burning, pyrolysis) or produce gaseous products (e.g. anaerobic digestion).

Potential Sources of Waste Inputs

The United States Environmental Protection Agency estimates that in 2006 there were 251 million tons of municipal solid waste, or 4.6 pounds generated per day per person in the USA. Much of this mass is considered unsuitable for oil conversion.

Limitations

The process only breaks long molecular chains into shorter ones, so small molecules such as carbon dioxide or methane cannot be converted to oil through this process. However, the methane in the feedstock is recovered and burned to heat the water that is an essential part of the process. In addition, the gas can be burned in a combined heat and power plant, consisting of a gas turbine which drives a generator to create electricity and a heat exchanger to heat the process input water from the exhaust gas. The electricity can be sold to the power grid, for example under a feed-in tariff scheme. This also increases the overall efficiency of the process (already said to be over 85% of feedstock energy content).

Another option is to sell the methane product as biogas. For example, biogas can be compressed, much like natural gas, and used to power motor vehicles.

Many agricultural and animal wastes could be processed, but many of these are already used as fertilizer, animal feed, and, in some cases, as feedstocks for paper mills or as boiler fuel. Energy crops constitute another potentially large feedstock for thermal depolymerization.

BIOMASS HEATING SYSTEM

Biomass heating systems generate heat from biomass. The systems fall under the categories of:

- Direct combustion,
- Gasification,
- Combined heat and power (CHP),
- Anaerobic digestion,
- Aerobic digestion.

Benefits of Biomass Heating

The use of biomass in heating systems is beneficial because it uses agricultural, forest, urban and industrial residues and waste to produce heat and/or electricity with less effect on the environment than fossil fuels. This type of energy production has a limited long-term effect on the environment because the carbon in biomass is part of the natural carbon cycle; while the carbon in fossil fuels is not, and permanently adds carbon to the environment when burned for fuel (carbon footprint). Historically, before the use of fossil fuels in significant quantities, biomass in the form of wood fuel provided most of humanity's heating.

Forest Health

Because forest based biomass is typically derived from wood that has lower commercial value, forest biomass is typically harvested as a byproduct of other timber harvest operations. Biomass heating provides markets for lower value wood, which enables healthy and profitable forest management. In New England as of 2017, one of the greatest threats to forest health is conversion from forest to agriculture and development. Harvard Forest scientists reported in 2017, that 65 acres of forest were being lost per day by conversion. By providing markets for low grade wood, the value of forests is enhanced, which makes conversion to housing or agriculture less likely.

Drawbacks of Biomass Heating

On a large scale, the use of agricultural biomass removes agricultural land from food production, reduces the carbon sequestration capacity of forests that are not managed sustainably, and extracts nutrients from the soil. Combustion of biomass creates air pollutants and adds significant quantities of carbon to the atmosphere that may not be returned to the soil for many decades. The time delay between when biomass is burned and the time when carbon is pulled from the atmosphere as a plant or tree grows to replace it is known as carbon debt. The concept of carbon debt is subject to debate. Actual carbon impacts may be subject to philosophy, scale of harvest, land type, biomass type (grass, maize, new wood, waste wood, algae, for example), soil type, and other factors.

Using biomass as a fuel produces air pollution in the form of carbon monoxide, NOx (nitrogen oxides), VOCs (volatile organic compounds), particulates and other pollutants, in some cases at levels above those from traditional fuel sources such as coal or natural gas. Black carbon – a pollutant created by incomplete combustion of fossil fuels, biofuels, and biomass – is possibly the second largest contributor to global warming. In 2009 a Swedish study of the giant brown haze that periodically covers large areas in South Asia determined that it had been principally produced by biomass burning, and to a lesser extent by fossil-fuel burning. Researchers measured a significant concentration of ^{14}C, which is associated with recent plant life rather than with fossil fuels. Modern biomass burning appliances dramatically reduce harmful emissions with advanced technology such as oxygen trim systems.

On combustion, the carbon from biomass is released into the atmosphere as carbon dioxide (CO_2). The amount of carbon stored in dry wood is approximately 50% by weight. When from agricultural sources, plant matter used as a fuel can be replaced by planting for new growth. When the biomass is from forests, the time to recapture the carbon stored is generally longer, and the carbon storage capacity of the forest may be reduced overall if destructive forestry techniques are employed.

The forest biomass-is-carbon-neutral proposal put forward in the early 1990s has been superseded by more recent science that recognizes that mature, intact forests sequester carbon more effectively than cut-over areas. When a tree's carbon is released into the atmosphere in a single pulse, it contributes to climate change much more than woodland timber rotting slowly over decades. Some studies indicate that "even after 50 years the forest has not recovered to its initial carbon storage" and "the optimal strategy is likely to be protection of the standing forest". Other studies show that carbon storage is dependent upon the forest and the use of the harvested biomass. Forests are often managed for multiple aged trees with more frequent, smaller harvests of mature trees. These forests interact with carbon differently than mature forests that are clear-cut. Also, the more efficient the conversion of wood to energy, the less wood that is used and shorter the carbon cycle will be.

Biomass Heating in our World

The oil price increases since 2003 and consequent price increases for natural gas and coal have increased the value of biomass for heat generation. Forest renderings, agricultural waste, and crops grown specifically for energy production become competitive as the prices of energy dense fossil fuels rise. Efforts to develop this potential may have the effect of regenerating mismanaged croplands and be a cog in the wheel of a decentralized, multi-dimensional renewable energy industry. Efforts to promote and advance these methods became common throughout the European Union through the 2000s. In other areas of the world, inefficient and polluting means to generate heat from biomass coupled with poor forest practices have significantly added to environmental degradation.

Biomass heating system for one building complex in the Spanish Basque Country.

Buffer Tanks

Buffer tanks store the hot water the biomass appliance generates and circulates it around the heating system. Sometimes referred to as 'thermal stores', they are crucial for the efficient operation of all biomass boilers where the system loading fluctuates rapidly, or the volume of water in the complete hydraulic system is relatively small. Using a suitably sized buffer vessel prevents rapid cycling of the boiler when the loading is below the minimum boiler output. Rapid cycling of the boiler causes a large increase in harmful emissions such as Carbon monoxide, dust, and NOx, greatly reduces boiler efficiency and increases electrical consumption of the unit. In addition, service and maintenance requirements will be increased as parts are stressed by rapid heating and cooling cycles. Although most boilers claim to be able to turn down to 30% of nominal output, in the real world this is often not achievable due to differences in the fuel from the 'ideal' or test fuel. A suitably sized buffer tank should therefore be considered where the loading of the boiler drops below 50% of the nominal output – in other words unless the biomass component is purely base load, the system should include a buffer tank. In any case where the secondary system does not contain sufficient water for safe removal of residual heat from the biomass boiler irrespective of the loading conditions, the system must include a suitably sized buffer tank. The residual heat from a biomass unit varies greatly depending on the boiler design and the thermal mass of the combustion chamber. Light weight, fast response boilers require only 10L/kW, while industrial wet wood units with very high thermal mass require 40L/kW.

Types of Biomass Heating Systems

The use of Biomass in heating systems has a use in many different types of buildings, and all have different uses. There are four main types of heating systems that use biomass to heat a boiler. The types are Fully Automated, Semi-Automated, Pellet-Fired, and Combined Heat and Power.

Biomass heating plant in Austria; the heat power is about 1000 kW.

Fully Automated

In fully automated systems chipped or ground up waste wood is brought to the site by delivery trucks and dropped into a holding tank. A system of conveyors then transports the wood from the holding tank to the boiler at a certain managed rate. This rate is managed by computer controls and a laser that measures the load of fuel the conveyor is bringing in. The system automatically goes on and off to maintain the pressure and temperature within the boiler. Fully automated systems offer a great deal of ease in their operation because they only require the operator of the system to control the computer, and not the transport of wood while offering comprehensive and cost effective solutions to complex industrial challenges.

Semi-automated or "Surge Bin"

Semi-automated or "Surge Bin" systems are very similar to fully automated systems except they require more manpower to keep operational. They have smaller holding tanks, and a much simpler conveyor systems which will require personnel to maintain the systems operation. The reasoning for the changes from the fully automated system is the efficiency of the system. The heat created by the combustor can be used to directly heat the air or it can be used to heat water in a boiler system which acts as the medium by which the heat is delivered. Wood fire fueled boilers are most efficient when they are running at their highest capacity, and the heat required most days of the year will not be the peak heat requirement for the year. Considering that the system will only need to run at a high capacity a few days of the year, it is made to meet the requirements for the majority of the year to maintain its high efficiency.

Pellet-fired

The third main type of biomass heating systems are pellet-fired systems. Pellets are a processed form of wood, which make them more expensive. Although they are more expensive, they are much more condensed and uniform, and therefore are more efficient. Further, it is relatively easy to automatically feed pellets to boilers. In these systems, the pellets are stored in a grain-type storage silo, and gravity is used to move them to the boiler. The storage requirements are much smaller for pellet-fired systems because of their condensed nature, which also helps cut down costs. these systems are used for a wide variety of facilities, but they are most efficient and cost effective for places where space for storage and conveyor systems is limited, and where the pellets are made fairly close to the facility.

Agricultural Pellet Systems

One subcategory of pellet systems are boilers or burners capable of burning pellet with higher ash rate (paper pellets, hay pellets, straw pellets). One of this kind is PETROJET pellet burner with rotating cylindrical burning chamber. In terms of efficiencies advanced pellet boilers can exceed other forms of biomass because of the more stable fuel charataristics. Advanced pellet boilers can even work in condensing mode and cool down combustion gases to 30-40 °C, instead of 120 °C before sent into the flue.

Combined Heat and Power

Combined heat and power systems are very useful systems in which wood waste, such as wood chips, is used to generate power, and heat is created as a byproduct of the power generation system. They have a very high cost because of the high pressure operation. Because of this high pressure operation, the need for a highly trained operator is mandatory, and will raise the cost of operation. Another drawback is that while they produce electricity they will produce heat, and if producing heat is not desirable for certain parts of the year, the addition of a cooling tower is necessary, and will also raise the cost.

There are certain situations where CHP is a good option. Wood product manufacturers would use a combined heat and power system because they have a large supply of waste wood, and a need for both heat and power. Other places where these systems would be optimal are hospitals and prisons, which need energy, and heat for hot water. These systems are sized so that they will produce enough heat to match the average heat load so that no additional heat is needed, and a cooling tower is not needed.

References

- Johnson, robin (2007). "torrefaction - a warmer solution to a colder climate". World conservation and wildlife trust. Retrieved september 30, 2013
- "Pyrolysis". Compendium of chemical terminology. International union of pure and applied chemistry. 2009. P. 1824. Doi:10.1351/goldbook.p04961. Isbn 978-0-9678550-9-7. Retrieved 2018-01-10
- Gasification: studentenergy.org, Retrieved 12 April, 2019
- Austin, anna (april 20, 2010). "french torrefaction firm targets north america". Biomass power and thermal. Retrieved february 29, 2012
- Gasification-explained: allpowerlabs.com, Retrieved 21 March, 2019
- "U.S. Product supplied for crude oil and petroleum products". Tonto.eia.doe.gov. Archived from the original on 28 february 2011. Retrieved 3 april 2018
- Moulijn, jacob a.; makkee, michiel; van diepen, annelies e. (may 2013). Chemical process technology. Wiley. Pp. 193–200. Isbn 978-1-4443-2025-1
- Fermentation: bioenergyconsult.com, Retrieved 1 March, 2019
- National non-food crops centre. "nnfcc renewable fuels and energy factsheet: anaerobic digestion", retrieved on 2011-11-22
- Fermentation, wiki-biomass: eubia.org, Retrieved 2 May, 2019
- "Biomass program, direct hydrothermal liquefaction". Us department of energy. Energy efficiency and renewable energy. 2005-10-13. Archived from the original on 2007-03-12. Retrieved 2008-01-12

Biofuel

CHAPTER 5

Biofuel is a fuel which is produced from biomass through modern processes, instead of geologic processes. The common types of biofuels include bioethanol, biohydrogen, biogas, biobutanol and biodiesel. This chapter has been carefully written to provide an easy understanding of these types of biofuels.

Biofuel is any fuel that is derived from biomass—that is, plant or algae material or animal waste. Since such feedstock material can be replenished readily, biofuel is considered to be a source of renewable energy, unlike fossil fuels such as petroleum, coal, and natural gas. Biofuel is commonly advocated as a cost-effective and environmentally benign alternative to petroleum and other fossil fuels, particularly within the context of rising petroleum prices and increased concern over the contributions made by fossil fuels to global warming. Many critics express concerns about the scope of the expansion of certain biofuels because of the economic and environmental costs associated with the refining process and the potential removal of vast areas of arable land from food production.

Ethanol gas fuel pump delivering the E85 mixture to an automobile.

Types of Biofuels

Some long-exploited biofuels, such as wood, can be used directly as a raw material that is burned to produce heat. The heat, in turn, can be used to run generators in a power plant to produce electricity. A number of existing power facilities burn grass, wood, or other kinds of biomass.

Liquid biofuels are of particular interest because of the vast infrastructure already in place to use them, especially for transportation. The liquid biofuel in greatest production is ethanol (ethyl alcohol), which is made by fermenting starch or sugar. Brazil and the United States are among the leading producers of ethanol. In the United States ethanol biofuel is made primarily from corn (maize) grain, and it is typically blended with gasoline to produce "gasohol," a fuel that is 10 percent ethanol. In Brazil, ethanol biofuel is made primarily from sugarcane, and it is commonly used as a 100-percent-ethanol fuel or in gasoline blends containing 85 percent ethanol. Unlike the "first-generation" ethanol biofuel produced from food crops, "second-generation" cellulosic ethanol is derived from low-value biomass that possesses a high cellulose content, including wood chips, crop residues, and municipal waste. Cellulosic ethanol is commonly made from sugarcane bagasse, a waste product from sugar processing, or from various grasses that can be cultivated on low-quality land. Given that the conversion rate is lower than with first-generation biofuels, cellulosic ethanol is dominantly used as a gasoline additive.

An ethanol production plant in South Dakota, U.S.

The second most common liquid biofuel is biodiesel, which is made primarily from oily plants (such as the soybean or oil palm) and to a lesser extent from other oily sources (such as waste cooking fat from restaurant deep-frying). Biodiesel, which has found greatest acceptance in Europe, is used in diesel engines and usually blended with petroleum diesel fuel in various percentages. The use of algae and cyanobacteria as a source of "third-generation" biodiesel holds promise but has been difficult to develop economically. Some algal species contain up to 40 percent lipids by weight, which can be converted into biodiesel or synthetic petroleum. Some estimates state that algae and cyanobacteria could yield between 10 and 100 times more fuel per unit area than second-generation biofuels.

Algal biofuel.

Other biofuels include methane gas and biogas—which can be derived from the decomposition of biomass in the absence of oxygen—and methanol, butanol, and dimethyl ether—which are in development.

Economic and Environmental Considerations

In evaluating the economic benefits of biofuels, the energy required to produce them has to be taken into account. For example, the process of growing corn to produce ethanol consumes fossil fuels in farming equipment, in fertilizer manufacturing, in corn transportation, and in ethanol distillation. In this respect, ethanol made from corn represents a relatively small energy gain; the energy gain from sugarcane is greater and that from cellulosic ethanol or algae biodiesel could be even greater.

Biofuels also supply environmental benefits but, depending on how they are manufactured, can also have serious environmental drawbacks. As a renewable energy source, plant-based biofuels in principle make little net contribution to global warming and climate change; the carbon dioxide (a major greenhouse gas) that enters the air during combustion will have been removed from the air earlier as growing plants engage in photosynthesis. Such a material is said to be "carbon neutral." In practice, however, the industrial production of agricultural biofuels can result in additional emissions of greenhouse gases that may offset the benefits of using a renewable fuel. These emissions include carbon dioxide from the burning of fossil fuels during the production process and nitrous oxide from soil that has been treated with nitrogen fertilizer. In this regard, cellulosic biomass is considered to be more beneficial.

Land use is also a major factor in evaluating the benefits of biofuels. The use of regular feedstock, such as corn and soybeans, as a primary component of first-generation biofuels sparked the "food versus fuel" debate. In diverting arable land and feedstock from the human food chain, biofuel production can affect the economics of food price and availability. In addition, energy crops grown for biofuel can compete for the world's natural habitats. For example, emphasis on ethanol derived from corn is shifting grasslands and brushlands to corn monocultures, and emphasis on biodiesel is bringing down ancient tropical forests to make way for oil palm plantations. Loss of natural habitat can change the hydrology, increase erosion, and generally reduce biodiversity of wildlife areas. The clearing of land can also result in the sudden release of a large amount of carbon dioxide as the plant matter that it contains is burned or allowed to decay.

Some of the disadvantages of biofuels apply mainly to low-diversity biofuel sources—corn, soybeans, sugarcane, oil palms—which are traditional agricultural crops. One alternative involves the use of highly diverse mixtures of species, with the North American tallgrass prairie as a specific example. Converting degraded agricultural land that is out of production to such high-diversity biofuel sources could increase wildlife area, reduce erosion, cleanse waterborne pollutants, store carbon dioxide from the air as carbon compounds in the soil, and ultimately restore fertility to degraded lands. Such biofuels could be burned directly to generate electricity or converted to liquid fuels as technologies develop.

The proper way to grow biofuels to serve all needs simultaneously will continue to be a matter of much experimentation and debate, but the fast growth in biofuel production will likely continue. In the United States the Energy Independence and Security Act of 2007 mandated the use of 136 billion litres (36 billion gallons) of biofuels annually by 2022, more than a sixfold increase over 2006 production levels. The legislation also requires, with certain stipulations, that 79 billion litres (21 billion gallons) of the total amount be biofuels other than corn-derived ethanol, and it continued certain government subsidies and tax incentives for biofuel production.

Biofuels testing centre: Workers at the biofuels testing centre at the National Renewable Energy Laboratory (NREL) in Golden.

One distinctive promise of biofuels is that, in combination with an emerging technology called carbon capture and storage, the process of producing and using biofuels may be capable of perpetually removing carbon dioxide from the atmosphere. Under this vision, biofuel crops would remove carbon dioxide from the air as they grow, and energy facilities would capture the carbon dioxide given off as biofuels are burned to generate power. Captured carbon dioxide could be sequestered (stored) in long-term repositories such as geologic formations beneath the land, in sediments of the deep ocean, or conceivably as solids such as carbonates.

Biofuel Uses

Many assume that the material is used just for transportation. But biofuel can provide hydrogen, clean up oil, work as cooking oil and more. Biofuels can work as an alternative to replacing energy needs from vehicle fuel to central home heating. The different uses of biofuel are:

Transportation

More than 30% of energy consumed in the United States is used for vehicle transportation. Across the globe, transport takes account of 24% of energy and more than 60% of absorbed oil. This means that over a third of oil is used to operate vehicles.

The main problem with alternatives is that solar, wind and other alternative power is not practical for transportation. Experts believe that efficient breakthroughs in practical technology advances are still decades away.

In short, biofuel can be turned into a hydrogen steam that is meant to be used in adjoining fuel-cell. More major car brands have already invested in stations for biofuel-powered vehicles.

Energy Generation

In addition to producing fuel for transportation, fuel cells have a power-generating application that is available for electricity. Biofuel can be used to generate power in backup systems where emission matters most. This includes facilities such as schools, hospitals and other forms located in residential areas.

In fact, the largest market for biofuel to turn into energy generation for over 350,000 homes from landfill gas in the United Kingdom.

Provide Heat

Bioheat has grown over the past few years. As the primary use of natural gas that comes from fossil fuel, the heat that comes from hydraulic fracturing will lead to the production of natural gas. While natural gas does not need to come from fossil material, it is also able to originate from the recently grown material.

A majority of biofuel that is used for heating is substantial. As wood is the most practical method to heat, houses that use wood burning stoves rather than gas or electricity. A blend of biodiesel will reduce the emission of both nitrogen and sulphur dioxide.

Charging Electronics

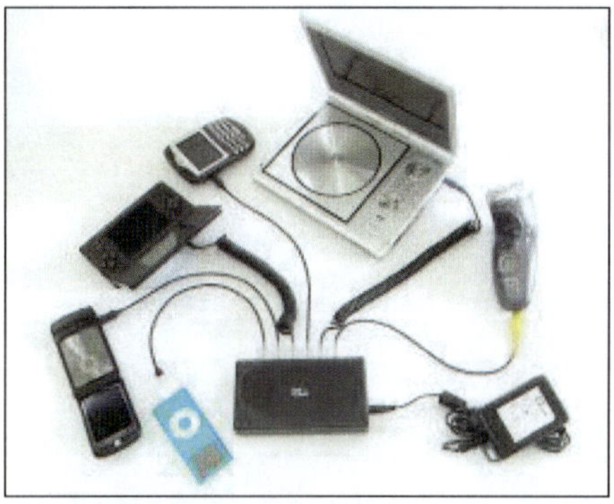

According to scientists from Saint Luis University, a fuel cell was developed with cooking oil and sugar to generate electricity; consumers will be able to use these cells instead of generating electricity. Consumers may be able to use fuel cells in place of batteries to charge anything from computers to cell phones. While they are still in the process of development, cells have the potential to become a ready source of power.

Clean Oil Spills and Grease

Biofuel is known to be environmentally-friendly, biofuel can also help to clean up oil spills and grease. It has been tested to work as a potential cleaning agent for areas where crude oil contaminated the waters.

The results have also been found to increase the recovery areas and allow it to be removed from the water. Biofuel can also be used as an industrial solvent for cleaning metal, which is also beneficial due to its lack of toxic impact.

Cooking

While kerosene is the most common ingredient to use for stoves and non-wick lanterns, biodiesel works just as great.

Lubricate

Diesel fuel is required to reduce the sulphur concentration as sulphur provides the most lubricity of fuel. This is important when it comes to keeping the engine properly functioning and to avoid premature infection failure.

Remove Paint and Adhesive

Biofuel can replace the toxic products that are designed to remove paint and adhesives. Biofuel is also considered as the best method for removing non-critical applications.

Create Energy when Fossil Fuel Runs Out

As the oil supply is starting to run out. This has caused us to question how fuel can be extracted without destroying the environment. Biofuel –will help the government create a stable method of producing energy that is cost-effective.

Reduce Cost and need for Imported Oil

More than 84% of the world's petroleum is used in the United States. Despite the increase in fuel demands, the U.S. has recently started to decrease the need since 2006. This allows biofuels to become the best factor in energy reduction.

Analysts say that replacing imported oil with biofuel will help to stabilise the economy when oil is disrupted. It does not matter how much the United States spends on oil import but how the overall economy must be stabilised.

Categories of Biofuels

- First generation biofuels:
    - These are made from food sources such as sugar, starch, vegetable oil, or animal fats using conventional technology.
    - Common first-generation biofuels include Bioalcohols, Biodiesel, Vegetable oil, Bioethers, Biogas.
    - Though the process of conversion is easy, but use of food sources in the production of biofuels creates an imbalance in food economy, leading to increased food prices and hunger.
- Second Generation Biofuels:
    - These are produced from non-food crops or portions of food crops that are not edible and considered as wastes, e.g. stems, husks, wood chips, and fruit skins and peeling.

- Thermochemical reactions or biochemical conversion process is used for producing such fuels.
- Examples include cellulose ethanol, biodiesel.
- Though these fuels do not affect food economy, their production is quite complicated.
- Also, it is reported that these biofuels emit less greenhouse gases when compared to first generation biofuels.

- Third Generation Biofuels:
  - These are produced from micro-organisms like algae.
  - Example - Butanol.
  - Micro-organisms like algae can be grown using land and water unsuitable for food production, therefore reducing the strain on already depleted water sources.
  - One disadvantage is that fertilizers used in the production of such crops lead to environment pollution.
- Fourth Generation Biofuels:
  - In the production of these fuels, crops that are genetically engineered to take in high amounts of carbon are grown and harvested as biomass.
  - The crops are then converted into fuel using second generation techniques.
  - The fuel is pre-combusted and the carbon is captured. Then the carbon is geo-sequestered, meaning that the carbon is stored in depleted oil or gas fields or in unmineable coal seams.
  - Some of these fuels are considered as carbon negative as their production pulls out carbon from environment.

BIOETHANOL

Ethanol produced from renewable energy sources is one of the most promising biofuels for the future. Although bioethanol fuels can be manufactured using the chemical reaction between ethylene and stream, it is mainly produced through fermentation of sugars derived from crops containing starch, such as corn, wheat, sugar cane, sorghum plants, etc.

It is currently used in the fuel industry as an additive for petrol. It is a high octane fuel and has replaced lead as an octane enhancer in petrol. Blending ethanol with petrol oxygenates the fuel mixture so that it burns completely and reduces harmful emissions. The most common blend is 90% petrol and 10% ethanol.

Bioethanol is entirely comprised of biological products, and hence the combustion of bioethanol results in cleaner emissions (carbon dioxide, steam and heat). Carbon dioxide is absorbed by plants and processed via photosynthesis to help the plant grow. This cycle of creation and energy combustion means bioethanol could potentially be a carbon neutral fuel source.

Bioethanol Production

Biomass wastes consist of a complex mixture of carbohydrate polymers such as cellulose, hemi cellulose and lignin. Biomass is pre-treated with acids or allowed to react with enzymes to reduce the size of the feedstock and produce sugars.

The carbohydrate polymers are broken down with the help of enzymes or dilute acids into sucrose sugar and then fermented into bioethanol. However, the lignin present in the biomass is used as a fuel for boilers in which bioethanol is produced. Enzymatic hydrolysis, concentrated acid hydrolysis and dilute acid hydrolysis are the three basic methods for extracting sugar from biomass.

Biomass can be used to produce large amounts of bioethanol.

On the other hand, corn can be processed into ethanol by either dry milling or wet milling. In the dry milling process, the corn kernel is cleaned and broken down into fine particles through hammer milling. This process creates a powder with a course flour-type consistency. The wet milling process involves soaking corn kernel in warm water to break down its proteins and release starch.

Once the biomass or corn is broken down into sugar via the hydrolysis process, the sugar solution is ready to be fermented into ethanol. The addition of yeast to the solution followed by rapid heating converts the sucrose sugars into fructose and glucose with the help of invertase present in the yeast.

These sugars further react with another enzyme, zymase, contained in the yeast to yield ethanol and carbon dioxide. The fermentation process is carried out at temperatures between 250 and 300 °C. The ethanol thus produced undergoes the fractional distillation process to remove excess water produced during the fermentation process.

Benefits of Bioethanol

Bioethanol has a number of benefits when compared to conventional fuels. Firstly, it is produced from a renewable resource (such as crops). There is therefore little/no net carbon dioxide added to the atmosphere, making bioethanol an environmentally beneficial energy source.

The road transport network contributes a great deal to the release of greenhouse gas emissions into the atmosphere, and with the use of bioethanol, emission rates can be drastically reduced. It is also biodegradable, and less toxic than fossil fuels.

In addition, blending bioethanol with petrol compensates for the diminishing oil supplies across the globe thereby ensuring higher fuel security and avoiding foreign reliance for fuel supply between countries. The rural community will also benefit from the increased demand to grow the necessary crops required for producing bioethanol.

It also reduces the emission of carbon monoxide produced by old vehicle engines and thus improves air quality. Another key benefit of bioethanol is the ease of integrating it with the existing road transport fuel system – bioethanol can be easily blended with conventional fuels (up to 15%) without any need for engine modifications.

Applications

Bioethanol can be used in petrol engines as an alternative for gasoline. It can be mixed with gasoline to virtually any percentage. Most of the existing petrol engines operate on blends of up to 15% bioethanol with petroleum.

The higher octane rating of bioethanol than ethanol-free gasoline increases an engine's compression ratio giving increased thermal efficiency. It is also used to fuel bioethanol fireplaces. It is extremely suitable for residential use as it is flueless and does not require a chimney.

Other major applications of bioethanol include the following:

- Fuel for power generation by thermal combustion.
- Fuel in cogeneration systems.
- Feedstock in the chemicals industry.
- Fuel for fuel cells by thermochemical reactions.

Environmental Impacts

Improving the quality of air is one of the most important functions of bioethanol. When added to fuel, bioethanol reduces the use of cancer-causing gasoline compounds such as ethylbenzene, xylene, toluene and benzene. It also reduces the emissions of small particulates and soot from motor fuels, and greenhouse gas emissions.

Water-saving ethanol plant designs are very common. In addition, the water discharged from these plants is regulated such that the water is environmentally neutral when it leaves the plant.

Certain plants manage and reuse the wastewater generated during the ethanol process. Therefore, it is evident that using bioethanol has a positive effect on ecology, minimizes exhaust gas emissions and improves energy safety and operation of transport facilities.

BIODIESEL

Biodiesel in simple words is an advanced form of Biofuel. It refers to an animal or vegetable fat based renewable fuel, made up of a long chain of chemical compounds like propyl, ethyl, methyl etc. Created by chemically treating vegetable oil or animal fat with alcohol generating compounds, biodiesel is the revolution that is helping several countries overcome their dependence on diesel. It can be operated in any diesel engine with little or no modification to the engine or the fuel system.

In other words, biodiesel is domestically produced, clean burning fuel that can be produced from waste vegetable oils, animal fats, or recycled restaurant grease for use in diesel vehicles. It is biodegradable, less toxic and produce less pollutants when burnt completely. It can be used in pure form (B100) or blended with petroleum diesel. Most of the common blend include B2 (20% biodiesel, 80% petrodiesel), B5 (5% biodiesel, 95% petrodiesel) or B20 (20% biodiesel, 80% petrodiesel).

Biodiesel has been found to have significant environmental benefits in terms of reduced emissions, greater energy independence, decreased global warming impacts and a positive impact on agriculture. It is low in sulfur, and has rapid biodegradability and high lubricity. Under specific conditions, used vegetable oils can be recycled as feedstock for producing biodiesel. This can in turn reduce the loss of used oils in the environment.

Biodiesel has much higher cetane rating than other lower sulfur diesel fuels. Biodiesel addition reduces wear of fuel system and increases the service life of the fuel injection equipment. However, variations in energy density of biodiesel are more dependent on the feedstock used.

Benefits of Biodiesel

Biodiesel has a number of beneficial properties. The main benefit of biodiesel is that it is 'carbon neutral'. This effect occurs due to the fact that the oil crop absorbs the CO2 released during fuel combustion for its growth. Another feature of biodiesel is that it is biodegradable, which means it can decompose as the result of natural agents such as bacteria. Biodiesel degrades at a rate four times faster than conventional diesel fuel. As a result, the cleanup would be easier. It also reduces the country's dependence on imported oil thereby increasing energy security. It also contributes to the lubricity of an engine.

It acts as a solvent and loosens deposits or other junk present in the engine that could possibly clog. As pure biodiesel does not leave any deposits, it results in increased engine life.

A detailed explanation would state that biodiesel, reduces unburned hydrocarbons, particulate matter and other gases as compared to petroleum diesel. In the United States of America, it is produced on a large scale in a bid to reduce dependency on other countries for oil needs. This had led to several green jobs being created and has also contributed to the economy.

It produces less carbon output and few pollutants than traditional petroleum products. When compared to petroleum diesel, biodiesel produces less soot (particulate matter), carbon monoxide, unburned hydrocarbons, and sulfur dioxide. Pollution from oil causes more air pollution and many people fall sick every day. With use of biodiesel, people have got huge health benefits as they need to spend less money on healthcare products.

As more crops are grown to produce biodiesel, it has created vast employment opportunities. Local people are employed in biofuel production plants and increase in demand for biodiesel may create more jobs. People who create crops to be used in biodiesel enjoy benefits as increase in demand for biodiesel leads to increase in demand for suitable biofuel crops.

The last few years have seen the upsurge of several vehicles and machines, with engines that run on biodiesel. One constantly hears and reads about the same in newspapers and automobile magazines. A large number of vehicles in the United States of America, come with options of biodiesel engines, talking volumes about the boom of this technology. As per the EPA, the biodiesel industry produced a mind bogging one billion gallons plus of biodiesel in the past year, 2012.

Bio degradation rates of biodiesel is another area of discussion. It degrades much faster than petrol and diesel. Some reports suggest in a period of 28 days. The areas of concern with the generation of biodiesel is a rut in the agricultural patterns for rearing biodiesel crops. Monotonous growth of the same crop will affect the ecology and the quality of the land.

The debate on bio fuels still on and its effects are still being accessed. But that does not dampen the ambitions of Biodiesel. Biodiesel is fast catching up in other countries too, as they snow ball in to an economically competitive alternative for energy. With Biodiesel making it big, the companies that manufacture it are acting increasingly as power companies. With the sky-rocketing prices of oil and petroleum, several investors are moving towards investing in Bio fuels citing greener pastures and a better future on this side.

Biodiesel has several other key benefits that include the following:

- It is environment-friendly.
- It is safer than conventional diesel.
- It is non-toxic.
- It has higher flashpoint than conventional diesel.

Biodiesel Production

Biodiesel production is the process of producing the biofuel, biodiesel, through the chemical reactions of transesterification and esterification. This involves vegetable or animal fats and oils being reacted with short-chain alcohols (typically methanol or ethanol). The alcohols used should be of low molecular weight. Ethanol is the most used because of its low cost, however, greater conversions into biodiesel can be reached using methanol. Although the transesterification reaction can be catalyzed by either acids or bases, the base-catalyzed reaction is more common. This path has lower reaction times and catalyst cost than those acid catalysis. However, alkaline catalysis has the disadvantage of high sensitivity to both water and free fatty acids present in the oils.

Process Steps

The major steps required to synthesize biodiesel are as follows:

Feedstock Pretreatment

Common feedstock used in biodiesel production include yellow grease (recycled vegetable oil), "virgin" vegetable oil, and tallow. Recycled oil is processed to remove impurities from cooking, storage, and handling, such as dirt, charred food, and water. Virgin oils are refined, but not to a food-grade level. Degumming to remove phospholipids and other plant matter is common, though refinement processes vary. Water is removed as its presence during base-catalyzed transesterification will cause the triglycerides to hydrolyze, producing soap instead of biodiesel.

A sample of the cleaned feedstock is then tested, via titration against a standardized base solution, in order to determine the concentration of free fatty acids present in the vegetable oil sample. These acids are then either removed, typically through neutralization, or esterified, into biodiesel or glycerides.

Reactions

Base-catalyzed transesterification reacts lipids (fats and oils) with alcohol (typically methanol or ethanol) to produce biodiesel and an impure coproduct, glycerol. If the feedstock oil is used or has a high acid content, acid-catalyzed esterification can be used to react fatty acids with alcohol to produce biodiesel. Other methods, such as fixed-bed reactors, supercritical reactors, and ultrasonic reactors, forgo or decrease the use of chemical catalysts.

Product Purification

Products of the reaction include not only biodiesel, but also byproducts, soap, glycerol, excess alcohol, and trace amounts of water. All of these byproducts must be removed to meet the standards, but the order of removal is process-dependent.

The density of glycerol is greater than that of biodiesel, and this property difference is exploited to separate the bulk of the glycerol coproduct. Residual methanol is typically recovered by distillation and reused. Soaps can be removed or converted into acids. Residual water is also removed from the fuel.

Reactions

Transesterification

Animal and plant fats and oils are composed of triglycerides, which are esters formed by the reactions of three free fatty acids and the trihydric alcohol, glycerol. In the transesterification process, the added alcohol (commonly, methanol or ethanol) is deprotonated with a base to make it a stronger nucleophile.the reaction has no other inputs than the triglyceride and the alcohol. Under normal conditions, this reaction will proceed either exceedingly slowly or not at all, so heat, as well as catalysts (acid and/or base) are used to speed the reaction. It is important to note that the acid or base are not consumed by the transesterification reaction, thus they are not reactants, but catalysts. Common catalysts for transesterification include sodium hydroxide, potassium hydroxide, and sodium methoxide.

Almost all biodiesel is produced from virgin vegetable oils using the base-catalyzed technique as it is the most economical process for treating virgin vegetable oils, requiring only low temperatures and pressures and producing over 98% conversion yield (provided the starting oil is low in moisture and free fatty acids). However, biodiesel produced from other sources or by other methods may require acid catalysis, which is much slower. Since it is the predominant method for commercial-scale production, only the base-catalyzed transesterification process will be described.

Triglycerides (1) are reacted with an alcohol such as ethanol (2) to give ethyl esters of fatty acids (3) and glycerol (4):

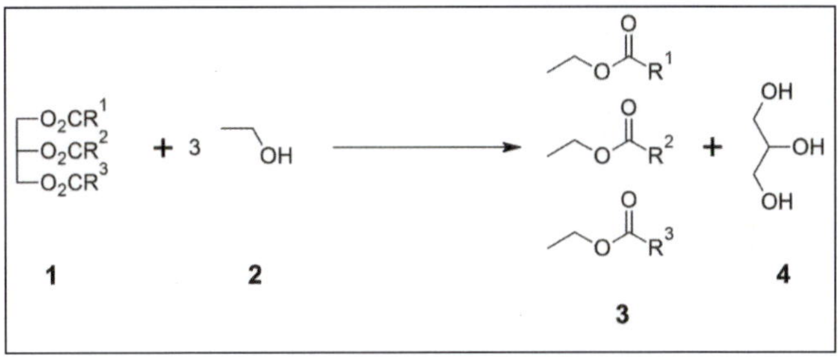

R^1, R^2, R^3 : Alkyl group.

The alcohol reacts with the fatty acids to form the mono-alkyl ester (biodiesel) and crude glycerol. The reaction between the biolipid (fat or oil) and the alcohol is a reversible reaction so excess alcohol must be added to ensure complete conversion.

Base-catalysed Transesterification Mechanism

The transesterification reaction is base catalyzed. Any strong base capable of deprotonating the alcohol will do (e.g. NaOH, KOH, sodium methoxide, etc.), but the sodium and potassium hydroxides are often chosen for their cost. The presence of water causes undesirable base hydrolysis, so the reaction must be kept dry.

In the transesterification mechanism, the carbonyl carbon of the starting ester ($RCOOR^1$) undergoes nucleophilic attack by the incoming alkoxide (R^2O^-) to give a tetrahedral intermediate, which either reverts to the starting material, or proceeds to the transesterified product ($RCOOR^2$). The various species exist in equilibrium, and the product distribution depends on the relative energies of the reactant and product.

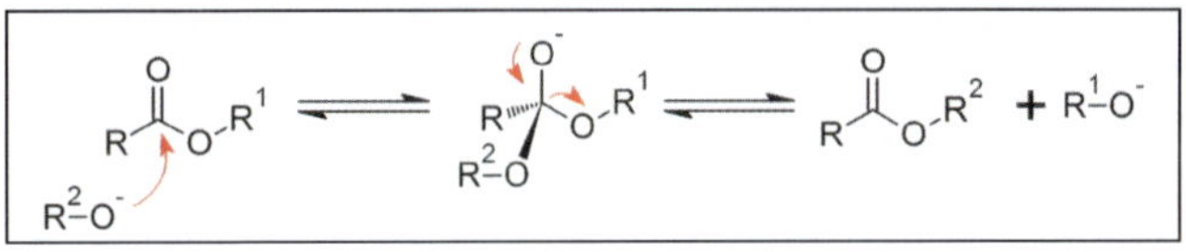

Production Methods

Supercritical Process

An alternative, catalyst-free method for transesterification uses supercritical methanol at high temperatures and pressures in a continuous process. In the supercritical state, the oil and methanol are in a single phase, and reaction occurs spontaneously and rapidly. The process can tolerate water in the feedstock, free fatty acids are converted to methyl esters instead of soap, so a wide variety of feedstocks can be used. Also the catalyst removal step is eliminated. High temperatures and pressures are required, but energy costs of production are similar or less than catalytic production routes.

Ultra- and High-shear in-line and Batch Reactors

Ultra- and High Shear in-line or batch reactors allow production of biodiesel continuously, semi- continuously, and in batch-mode. This drastically reduces production time and increases production volume.

The reaction takes place in the high-energetic shear zone of the Ultra- and High Shear mixer by reducing the droplet size of the immiscible liquids such as oil or fats and methanol. Therefore, the smaller the droplet size the larger the surface area the faster the catalyst can react.

Ultrasonic Reactor Method

In the ultrasonic reactor method, the ultrasonic waves cause the reaction mixture to produce and collapse bubbles constantly. This cavitation simultaneously provides the mixing and heating required to carry out the transesterification process. Thus, using an ultrasonic reactor for biodiesel production drastically reduces the reaction time, reaction temperatures, and energy input. Hence the process of transesterification can run inline rather than using the time consuming batch processing. Industrial scale ultrasonic devices allow for the industrial scale processing of several thousand barrels per day.

Lipase-catalyzed Method

Large amounts of research have focused recently on the use of enzymes as a catalyst for the transesterification. Researchers have found that very good yields could be obtained from crude and used oils using lipases. The use of lipases makes the reaction less sensitive to high free fatty-acid content, which is a problem with the standard biodiesel process. One problem with the lipase reaction is that methanol cannot be used because it inactivates the lipase catalyst after one batch. However, if methyl acetate is used instead of methanol, the lipase is not in-activated and can be used for several batches, making the lipase system much more cost effective.

Volatile Fatty Acids from Anaerobic Digestion of Waste Streams

Lipids have been drawing considerable attention as a substrate for biodiesel production owing to its sustainability, non-toxicity and energy efficient properties. However, due to cost reasons, attention must be focused on the non-edible sources of lipids, in particular oleaginous microorganisms. Such microbes have the ability to assimilate the carbon sources from a medium and convert the carbon into lipid storage materials. The lipids accumulated by these oleaginous cells can then be transesterified to form biodiesel.

BIOGAS

Biogas is the mixture of gases produced by the breakdown of organic matter in the absence of oxygen. Biogas can be produced from raw materials such as agricultural waste, manure, municipal waste, plant material, sewage, green waste or food waste. Biogas is a renewable energy source.

Biogas is produced by anaerobic digestion with methanogen or anaerobic organisms, which digest material inside a closed system, or fermentation of biodegradable materials. This closed system is called an anaerobic digester, biodigester or a bioreactor.

Biogas is primarily methane (CH_4) and carbon dioxide (CO_2) and may have small amounts of hydrogen sulfide (H_2S), moisture and siloxanes. The gases methane, hydrogen, and carbon monoxide (CO) can be combusted or oxidized with oxygen. This energy release allows biogas to be used as a fuel; it can be used for any heating purpose, such as cooking. It can also be used in a gas engine to convert the energy in the gas into electricity and heat.

Biogas can be compressed, the same way as natural gas is compressed to CNG, and used to power motor vehicles. In the United Kingdom, for example, biogas is estimated to have the potential to replace around 17% of vehicle fuel. It qualifies for renewable energy subsidies in some parts of the world. Biogas can be cleaned and upgraded to natural gas standards, when it becomes bio-methane. Biogas is considered to be a renewable resource because its production-and-use cycle is continuous, and it generates no net carbon dioxide. As the organic material grows, it is converted and used. It then regrows in a continually repeating cycle. From a carbon perspective, as much carbon dioxide is absorbed from the atmosphere in the growth of the primary bio-resource as is released, when the material is ultimately converted to energy.

Production

The biogas is a renewable energy that can be used for heating, electricity, and many other operations that use a reciprocating internal combustion engine, such as GE Jenbacher or Caterpillar gas engines. To provide these internal combustion engines with biogas having ample gas pressure to optimize combustion, within the European Union ATEX centrifugal fan units built in accordance with the European directive 2014/34/EU (previously 94/9/EG) are obligatory. These centrifugal fan units, for example Combimac, Meidinger AG or Witt & Sohn AG are suitable for use in Zone 1 and 2.

Other internal combustion engines such as gas turbines are suitable for the conversion of biogas into both electricity and heat. The digestate is the remaining inorganic matter that was not transformed into biogas. It can be used as an agricultural fertiliser.

Biogas production in rural Germany.

Biogas is produced either:

- As landfill gas (LFG), which is produced by the breakdown of biodegradable waste inside a landfill due to chemical reactions and microbes,
- As digested gas, produced inside an anaerobic digester.

Projects such as NANOCLEAN are nowadays developing new ways to produce biogas more efficiently, using iron oxide nanoparticles in the processes of organic waste treatment. This process can triple the production of biogas.

Biogas Plants

A *biogas plant* is the name often given to an anaerobic digester that treats farm wastes or energy crops. It can be produced using anaerobic digesters (air-tight tanks with different configurations). These plants can be fed with energy crops such as maize silage or biodegradable wastes including sewage sludge and food waste. During the process, the micro-organisms transform biomass waste into biogas (mainly methane and carbon dioxide) and digestate. Higher quantity of biogas could be produced when the wastewater is co-fermented with other residual from dairy industry, sugar industry, brewery industry. For example, while mixing 90% of wastewater from beer factory with 10% cow whey, the production of biogas is increased by 2.5 times compared to the biogas produced by wastewater from beer factory only.

Key Processes

There are two key processes: mesophilic and thermophilic digestion which is dependent on temperature. In experimental work at University of Alaska Fairbanks, a 1000-litre digester using psychrophiles harvested from "mud from a frozen lake in Alaska" has produced 200–300 liters of methane per day, about 20%–30% of the output from digesters in warmer climates.

Biogas can be produced from a vast variety of raw materials (feedstocks). The biggest role in the biogas production process is played by microbes feeding on the biomass.

Digestion carried out by these microorganisms creates methane, which can be used as it is locally or upgraded to biogas equivalent to natural gas quality, enabling the transport of the biogas over longer distances. Material containing organic nutrients is also produced in the process, and this can be utilized for purposes such as agriculture.

Stages in Biogas Production

Biogas is produced using well-established technology in a process involving several stages:

- Biowaste is crushed into smaller pieces and slurrified to prepare it for the anaerobic digestion process. Slurrifying means adding liquid to the biowaste to make it easier to process.
- Microbes need warm conditions, so the biowaste is heated to around 37 °C.
- The actual biogas production takes place through anaerobic digestion in large tanks for about three weeks.
- In the final stage, the gas is purified (upgraded) by removing impurities and carbon dioxide.

After this, the biogas is ready for use by enterprises and consumers, for example in a liquefied form or following injection into the gas pipeline network.

Turning Diverse Range of Materials into Gas

Biogas production starts from the arrival of feedstocks at the biogas plant. A diverse range of solid as well as sludge-like feedstocks can be used.

Materials suitable for biogas production include:

- Biodegradable waste from enterprises and industrial facilities, such as surplus lactose from the production of lactose-free dairy products.
- Spoiled food from shops.
- Biowaste generated by consumers.
- Sludge from wastewater treatment plants.
- Manure and field biomass from agriculture.

The material is typically delivered to the biogas plant's reception pit by lorry or waste management vehicle.

A delivery of solid matter such as biowaste will next undergo crushing to make its consistency as even as possible. At this point, water containing nutrients obtained from a further stage in the production process is also mixed with the feedstock to take the rate of solid matter down to only around one-tenth of the total volume.

This is also when any unwanted non-biodegradable waste, such as packaging plastic of out-of-date food waste from shops, is separated from the mixture. This waste is taken to a waste treatment facility where it is used to generate heat and electricity. Biomass that has passed through slurrification is combined with biomass delivered in the form of slurry to the biogas plant and pumped into the pre-digester tank where enzymes secreted by bacteria break down the biomass into an even finer consistency.

Next, the biomass is sanitized before entering the actual biogas reactor (digester). In sanitization, any harmful bacteria found in the material are eliminated by heating the mixture to above 70 °C for one hour. Once sanitized, the mass is pumped into the main reactor where biogas production takes place. Sanitization makes it possible to use the fertilizer product in agriculture.

Biomass is Turned into Gas by Microbes

In the biogas reactor, microbial action begins and the biomass enters a gradual process of fermentation.

In practice this means that microbes feed on the organic matter, such as proteins, carbohydrates and lipids, and their digestion turns these into methane and carbon dioxide.

Most of the organic matter is broken down into biogas – a mixture of methane and carbon dioxide – in approximately three weeks. The biogas is collected in a spherical gas holder from the top of the biogas reactors.

Digestate Utilized as Fertilizers or Gardening Soil

The residual solids and liquids created in biogas production are referred to as digestate. This digestate goes into a post-digester reactor and from there further into storage tanks. Digestates are well suited for uses such as fertilization of fields.

Digestates can also be centrifuged to separate the solid and liquid parts.

Solid digestates have uses such as fertilizers in agriculture or in landscaping and can also be turned into gardening soil through a process of maturation involving composting.

Digestates are centrifuged to yield enough process water for the slurrification of biowaste at the beginning of the process. This helps reduce the use of clean water. The centrifuged liquid is rich in nutrients, particularly nitrogen, that can be separated further using methods such as stripping technology and used as fertilizers or nutrient sources in industrial processes.

Clean Biogas Helps move towards Low-carbon Society

Gas would already be ready for several uses straight from the biogas plant gas holder. However, before being injected into the gas pipeline network or used to fuel vehicles, it will still undergo purification.

In this upgrading process, gas is filtered and flown into columns where it is scrubbed by cascading water at a very specific pressure and temperature. Water efficiently absorbs carbon dioxide and sulfur compounds contained by the gas. Biogas can also be purified using other methods, such as passing it through activated carbon filters to remove impurities.

The final upgraded biogas injected into the gas network is at least 95% and usually around 98% methane. Upgraded biogas still contains a couple of per cent of carbon dioxide as its further separation from methane is not cost-effective let alone sensible as regards the usability of the gas. Biogas is dried carefully before injection into the gas network to prevent condensation in winter subzero conditions.

The biogas produced can be used for purposes such as fuelling municipal waste management vehicles, urban buses or private cars. At the same time, gas serves as evidence of those practical actions that are taking us towards the low-carbon society of the future.

Landfill Gas

Landfill gas is produced by wet organic waste decomposing under anaerobic conditions in a biogas.

The waste is covered and mechanically compressed by the weight of the material that is deposited above. This material prevents oxygen exposure thus allowing anaerobic microbes to thrive. Biogas builds up and is slowly released into the atmosphere if the site has not been engineered to capture the gas. Landfill gas released in an uncontrolled way can be hazardous since it can become explosive when it escapes from the landfill and mixes with oxygen. The lower explosive limit is 5% methane and the upper is 15% methane.

The methane in biogas is 28 times more potent a greenhouse gas than carbon dioxide. Therefore, uncontained landfill gas, which escapes into the atmosphere may significantly contribute to the effects of global warming. In addition, volatile organic compounds (VOCs) in landfill gas contribute to the formation of photochemical smog.

Technical

Biochemical oxygen demand (BOD) is a measure of the amount of oxygen required by aerobic micro-organisms to decompose the organic matter in a sample of material being used in the biodigester as well as the BOD for the liquid discharge allows for the calculation of the daily energy output from a biodigester.

Another term related to biodigesters is effluent dirtiness, which tells how much organic material there is per unit of biogas source. Typical units for this measure are in mg BOD/litre. As an example, effluent dirtiness can range between 800–1200 mg BOD/litre in Panama.

From 1 kg of decommissioned kitchen bio-waste, 0.45 m^3 of biogas can be obtained. The price for collecting biological waste from households is approximately €70 per ton.

Composition

Typical composition of Biogas		
Compound	Formula	%
Methane	CH_4	50–75
Carbon dioxide	CO_2	25–50
Nitrogen	N_2	0–10
Hydrogen	H_2	0–1
Hydrogen sulfide	H_2S	0.1 –0.5
Oxygen	O_2	0–0.5

The composition of biogas varies depending upon the substrate composition, as well as the conditions within the anaerobic reactor (temperature, pH, and substrate concentration). Landfill gas typically has methane concentrations around 50%. Advanced waste treatment technologies can produce biogas with 55%–75% methane, which for reactors with free liquids can be increased to 80%–90% methane using in-situ gas purification techniques. As produced, biogas contains water vapor. The fractional volume of water vapor is a function of biogas temperature; correction of measured gas volume for water vapour content and thermal expansion is easily done via simple mathematics which yields the standardized volume of dry biogas.

In some cases, biogas contains siloxanes. They are formed from the anaerobic decomposition of materials commonly found in soaps and detergents. During combustion of biogas containing siloxanes, silicon is released and can combine with free oxygen or other elements in the combustion gas. Deposits are formed containing mostly silica (SiO_2) or silicates (Si_xO_y) and can contain calcium, sulfur, zinc, phosphorus. Such *white mineral* deposits accumulate to a surface thickness of several millimeters and must be removed by chemical or mechanical means.

Practical and cost-effective technologies to remove siloxanes and other biogas contaminants are available.

For 1000 kg (wet weight) of input to a typical biodigester, total solids may be 30% of the wet weight while volatile suspended solids may be 90% of the total solids. Protein would be 20% of the volatile solids, carbohydrates would be 70% of the volatile solids, and finally fats would be 10% of the volatile solids.

Benefits of Manure Derived Biogas

High levels of methane are produced when manure is stored under anaerobic conditions. During storage and when manure has been applied to the land, nitrous oxide is also produced as a byproduct of the denitrification process. Nitrous oxide (N_2O) is 320 times more aggressive as a greenhouse gas than carbon dioxide and methane 25 times more than carbon dioxide. By converting cow manure into methane biogas via anaerobic digestion, the millions of cattle in the United States would be able to produce 100 billion kiloWatt hours of electricity, enough to power millions of homes across the United States. In fact, one cow can produce enough manure in one day to generate 3 kiloWatt hours of electricity; only 2.4 kiloWatt hours of electricity are needed to power a single 100-Watt light bulb for one day. Furthermore, by converting cattle manure into methane biogas instead of letting it decompose, global warming gases could be reduced by 99 million metric tons or 4%.

Applications

Biogas can be used for electricity production on sewage works, in a CHP gas engine, where the waste heat from the engine is conveniently used for heating the digester; cooking; space heating; water heating; and process heating. If compressed, it can replace compressed natural gas for use in vehicles, where it can fuel an internal combustion engine or fuel cells and is a much more effective displacer of carbon dioxide than the normal use in on-site CHP plants.

A biogas bus in Linköping, Sweden.

Biogas Upgrading

Raw biogas produced from digestion is roughly 60% methane and 39% CO_2 with trace elements of H_2S: inadequate for use in machinery. The corrosive nature of H_2S alone is enough to destroy the mechanisms.

Methane in biogas can be concentrated via a biogas upgrader to the same standards as fossil natural gas, which itself has to go through a cleaning process, and becomes *biomethane*. If the local gas network allows, the producer of the biogas may use their distribution networks. Gas must be very clean to reach pipeline quality and must be of the correct composition for the distribution network to accept. Carbon dioxide, water, hydrogen sulfide, and particulates must be removed if present.

There are four main methods of upgrading: water washing, pressure swing absorption, selexol absorption, and amine gas treating. In addition to these, the use of membrane separation technology for biogas upgrading is increasing, and there are already several plants operating in Europe and USA.

The most prevalent method is water washing where high pressure gas flows into a column where the carbon dioxide and other trace elements are scrubbed by cascading water running counter-flow to the gas. This arrangement could deliver 98% methane with manufacturers guaranteeing maximum 2% methane loss in the system. It takes roughly between 3% and 6% of the total energy output in gas to run a biogas upgrading system.

Biogas Gas-grid Injection

Gas-grid injection is the injection of biogas into the methane grid (natural gas grid). Until the breakthrough of micro combined heat and power two-thirds of all the energy produced by biogas power plants was lost (as heat). Using the grid to transport the gas to customers, the energy can be used for on-site generation, resulting in a reduction of losses in the transportation of energy. Typical energy losses in natural gas transmission systems range from 1% to 2%; in electricity transmission they range from 5% to 8%.

Before being injected in the gas grid, biogas passes a cleaning process, during which it is upgraded to natural gas quality. During the cleaning process trace components harmful to the gas grid and the final users are removed.

Biogas in Transport

If concentrated and compressed, it can be used in vehicle transportation. Compressed biogas is becoming widely used in Sweden, Switzerland, and Germany. A biogas-powered train, named Biogaståget Amanda (The Biogas Train Amanda), has been in service in Sweden since 2005. Biogas powers automobiles. In 1974, a British documentary film titled *Sweet as a Nut* detailed the biogas production process from pig manure and showed how it fueled a custom-adapted combustion engine. In 2007, an estimated 12,000 vehicles were being fueled with upgraded biogas worldwide, mostly in Europe.

“Biogaståget Amanda” (“Amanda the Biogas Train”) train near.

Biogas is part of the wet gas and condensing gas (or air) category that includes mist or fog in the gas stream. The mist or fog is predominately water vapor that condenses on the sides of pipes or stacks throughout the gas flow. Biogas environments include wastewater digesters, landfills, and animal feeding operations (covered livestock lagoons).

Ultrasonic flow meters are one of the few devices capable of measuring in a biogas atmosphere. Most of thermal flow meters are unable to provide reliable data because the moisture causes steady high flow readings and continuous flow spiking, although there are single-point insertion thermal mass flow meters capable of accurately monitoring biogas flows with minimal pressure drop. They can handle moisture variations that occur in the flow stream because of daily and seasonal temperature fluctuations, and account for the moisture in the flow stream to produce a dry gas value.

Biogas Generated Heat and Electricity

Biogas can be used as the fuel in the system of producing biogas from agricultural wastes and co-generating heat and electricity in a combined heat and power (CHP) plant. Unlike the other green energy such as wind and solar, the biogas can be quickly accessed on demand. The global warming potential can also be greatly reduced when using biogas as the fuel instead of fossil fuel.

However, the acidification and eutrophication potentials produced by biogas are 25 and 12 times higher respectively than fossil fuel alternatives. This impact can be reduced by using correct combination of feedstocks, covered storage for digesters and improved techniques for retrieving escaped material. Overall, the results still suggest that using biogas can lead to significant reduction in most impacts compared to fossil fuel alternative. The balance between environmental damage and green house gas emission should still be considered while implicating the system.

BIOBUTANOL

Biobutanol is a four-carbon alcohol produced by the fermentation of biomass. It has a long hydrocarbon chain which renders it fairly non-polar. The production of biobutanol can be carried out in ethanol production facilities. The primary use of biobutanol is a fuel in an internal combustion engine.

Its properties are similar to that of gasoline. Some gasoline-powered vehicles can even use biobutanol without being modified. It can be blended with gasoline in concentrations up to 11.5% by volume. However, it has a lower energy content, on average 10-20%, than that of gasoline, which is a major disadvantage of biobutanol.

Biobutanol exhibits the potential to reduce carbon emissions by 85% when compared to gasoline, thus making it a viable and suitable alternative to gasoline and gasoline-ethanol blended fuels.

Biobutanol Production Technologies

Biobutanol is mainly derived from the fermentation of sugars in organic feedstocks. The most common method of producing biobutanol is the fermentation of simple sugars in biomass feedstock.

Butanol is a by-product of this process in addition to ethanol and acetone. The process, known as Acetone Butanol Ethanol, uses the microbial species Clostridium acetobutylicum. However, this microbe can easily be poisoned by butanol when the alcohol concentration rises above approximately 2%.

Biobutanol can also be produced using Ralstonia eutropha H16. This process requires the input of electricity and carbon dioxide, and the use of an electro-bioreactor. It can be produced from algae or diatoms with the help of solar energy.

In recent years, petroleum prices have been steadily increasing yet fermentation remains an inefficient method of producing biobutanol due to the low biobutanol yield of the Clostridium species.

Manipulation of metabolic networks within bacterial species can provide higher yields of biobutanol. Optimization can also be accomplished by the transfer of specific genetic information to other unicellular species to achieve a higher biobutanol production rate.

Advantages of Biobutanol

At high concentrations, biobutanol be blended with conventional petrol rather than ethanol for use in unmodified engines. Experiments have also proved that biobutanol can be used in unmodified conventional engines at 100%. However, no manufacturers have guaranteed use of blends greater than 15%.

Biobutanol has a higher energy content than ethanol. With an energy content of about 105,000 BTUs/gallon, biobutanol is close to the energy content of gasoline, which is roughly 114,000 BTUs/gallon.

Less corrosive and explosive than ethanol, biobutanol is also less susceptible to separation in the presence of water than ethanol. It can be produced domestically from a variety of feedstocks, which can also help drive the economy via the generation of jobs.

Carbon dioxide captured by growing feedstocks minimizes overall greenhouse gas emissions by balancing carbon dioxide released from burning biobutanol. Environmental Protection Agency (EPA) test results show that biobutanol reduces hydrocarbon, carbon monoxide and nitrogen oxide emissions.

Applications of Biobutanol

There is now increasing interest in the use of biobutanol as a transport fuel. Unfortunately though no production vehicle is known to be approved by manufacturers for use with 100% butanol.

Biobutanol also shows promise as an industrial solvent and chemical feedstock. In addition to this possible other applications may include paints/coatings, resins, plasticizers, pharmaceuticals, food grade extractants, chemical intermediates and herbicides.

BIOHYDROGEN

Biohydrogen is H_2 that is produced biologically. Interest is high in this technology because H_2 is a clean fuel and can be readily produced from certain kinds of biomass. Many challenges characterize this technology, including those intrinsic to H_2, such as storage and transportation of a noncondensible gas. Hydrogen producing organisms are poisoned by O_2. Yields of H_2 are often low.

Biochemical Principles

The main reactions involve fermentation of sugars. Important reactions start with glucose, which is converted to acetic acid:

$$C_6H_{12}O_6 + 2\,H_2O \rightarrow 2\,CH_3CO_2H + 2\,CO_2 + 4\,H_2$$

A related reaction gives formate instead of carbon dioxide:

$$C_6H_{12}O_6 + 2\ H_2O \rightarrow 2\ CH_3CO_2H + 2\ HCO_2H + 2\ H_2$$

These reactions are exergonic by 216 and 209 kcal/mol, respectively.

H_2 production is catalyzed by two hydrogenases. One is called [FeFe]-hydrogenase; the other is called [NiFe]-hydrogenase. Many organisms express these enzymes. Notable examples are members of the genera Clostridium, Desulfovibrio, Ralstonia, and the pathogen *Helicobacter*. *E. coli* is the workhorse for genetic engineering of hydrogenases.

It has been estimated that 99% of all organisms utilize dihydrogen (H_2). Most of these species are microbes and their ability to use H_2 as a metabolite arises from the expression of H_2 metalloenzymes known as hydrogenases. Hydrogenases are sub-classified into three different types based on the active site metal content: iron-iron hydrogenase, nickel-iron hydrogenase, and iron hydrogenase.

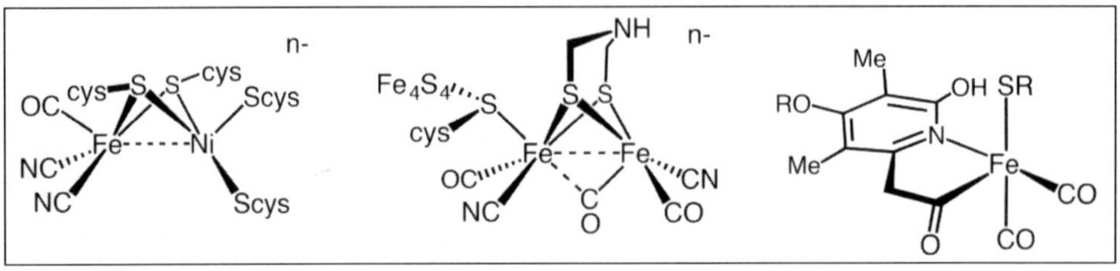

The active site structures of the three types of hydrogenase enzymes.

Production by Algae

The biological hydrogen production with algae is a method of photobiological water splitting which is done in a closed photobioreactor based on the production of hydrogen as a solar fuel by algae. Algae produce hydrogen under certain conditions. In 2000 it was discovered that if *C. reinhardtii* algae are deprived of sulfur they will switch from the production of oxygen, as in normal photosynthesis, to the production of hydrogen.

Photosynthesis

Photosynthesis in cyanobacteria and green algae splits water into hydrogen ions and electrons. The electrons are transported over ferredoxins. Fe-Fe-hydrogenases (enzymes) combine them into hydrogen gas. In *Chlamydomonas reinhardtii* Photosystem II produces in direct conversion of sunlight 80% of the electrons that end up in the hydrogen gas. Light-harvesting complex photosystem II light-harvesting protein LHCBM9 promotes efficient light energy dissipation. The Fe-Fe-hydrogenases need an anaerobic environment as they are inactivated by oxygen. Fourier transform infrared spectroscopy is used to examine metabolic pathways.

Specialized Chlorophyll

The chlorophyll (Chl) antenna size in green algae is minimized, or truncated, to maximize photobiological solar conversion efficiency and H_2 production. The truncated Chl antenna size minimizes absorption and wasteful dissipation of sunlight by individual cells, resulting in better light utilization efficiency and greater photosynthetic productivity by the green alga mass culture.

Economics

It would take about 25,000 square kilometre algal farming to produce biohydrogen equivalent to the energy provided by gasoline in the US alone. This area represents approximately 10% of the area devoted to growing soya in the US.

Bioreactor Design Issues

- Restriction of photosynthetic hydrogen production by accumulation of a proton gradient.
- Competitive inhibition of photosynthetic hydrogen production by carbon dioxide.
- Requirement for bicarbonate binding at photosystem II (PSII) for efficient photosynthetic activity.
- Competitive drainage of electrons by oxygen in algal hydrogen production.
- Economics must reach competitive price to other sources of energy and the economics are dependent on several parameters.
- A major technical obstacle is the efficiency in converting solar energy into chemical energy stored in molecular hydrogen.

Attempts are in progress to solve these problems via bioengineering.

BIOMASS BRIQUETTES

Biomass briquettes are a biofuel substitute to coal and charcoal. Briquettes are mostly used in the developing world, where cooking fuels are not as easily available. There has been a move to the use of briquettes in the developed world, where they are used to heat industrial boilers in order to produce electricity from steam. The briquettes are cofired with coal in order to create the heat supplied to the boiler.

Composition and Production

Biomass briquettes, mostly made of green waste and other organic materials, are commonly used for electricity generation, heat, and cooking fuel. These compressed compounds contain various organic materials, including rice husk, bagasse, ground nut shells, municipal solid waste, agricultural waste. The composition of the briquettes varies by area due to the availability of raw materials. The raw materials are gathered and compressed into briquette in order to burn longer and make transportation of the goods easier. These briquettes are very different from charcoal because they do not have large concentrations of carbonaceous substances and added materials. Compared to fossil fuels, the briquettes produce low net total greenhouse gas emissions because the materials used are already a part of the carbon cycle.

One of the most common variables of the biomass briquette production process is the way the biomass is dried out. Manufacturers can use torrefaction, carbonization, or varying degrees of pyrolysis. Researchers concluded that torrefaction and carbonization are the most efficient forms of drying out biomass, but the use of the briquette determines which method should be used.

Compaction is another factor affecting production. Some materials burn more efficiently if compacted at low pressures, such as corn stover grind. Other materials such as wheat and barley-straw require high amounts of pressure to produce heat. There are also different press technologies that can be used. A piston press is used to create solid briquettes for a wide array of purposes. Screw extrusion is used to compact biomass into loose, homogeneous briquettes that are substituted for coal in cofiring. This technology creates a toroidal, or doughnut-like, briquette. The hole in the center of the briquette allows for a larger surface area, creating a higher combustion rate.

Cofiring

Cofiring relates to the combustion of two different types of materials. The process is primarily used to decrease CO_2 emissions despite the resulting lower energy efficiency and higher variable cost. The combination of materials usually contains a high carbon emitting substance such as coal and a lesser CO_2 emitting material such as biomass. Even though CO_2 will still be emitted through the combustion of biomass, the net carbon emitted is nearly negligible. This is due to the fact that the material gathered for the composition of the briquettes are still contained in the carbon cycle whereas fossil fuel combustion releases CO_2 that has been sequestered for millennia. Boilers in power plants are traditionally heated by the combustion of coal, but if cofiring were to be implemented, then the CO_2 emissions would decrease while still maintaining the heat inputted to the boiler. Implementing cofiring would require few modifications to the current characteristics to power plants, as only the fuel for the boiler would be altered. A moderate investment would be required for implementing biomass briquettes into the combustion process.

Cofiring is considered the most cost-efficient means of biomass. A higher combustion rate will occur when cofiring is implemented in a boiler when compared to burning only biomass. The compressed biomass is also much easier to transport since it is more dense, therefore allowing more biomass to be transported per shipment when compared to loose biomass. Some sources agree that a near-term solution for the greenhouse gas emission problem may lie in cofiring.

Compared to Coal

The use of biomass briquettes has been steadily increasing as industries realize the benefits of decreasing pollution through the use of biomass briquettes. Briquettes provide higher calorific value per dollar than coal when used for firing industrial boilers. Along with higher calorific value, biomass briquettes on average saved 30–40% of boiler fuel cost. But other sources suggest that cofiring is more expensive due to the widespread availability of coal and its low cost. However, in the long run, briquettes can only limit the use of coal to a small extent, but it is increasingly being pursued by industries and factories all over the world. Both raw materials can be produced or mined domestically in the United States, creating a fuel source that is free from foreign dependence and less polluting than raw fossil fuel incineration.

Environmentally, the use of biomass briquettes produces much fewer greenhouse gases, specifically, 13.8% to 41.7% CO_2 and NO_X. There was also a reduction from 11.1% to 38.5% in SO_2 emissions when compared to coal from three different leading producers, EKCC Coal, Decanter Coal, and Alden Coal. Biomass briquettes are also fairly resistant to water degradation, an improvement over the difficulties encountered with the burning of wet coal. However, the briquettes are best used only as a supplement to coal. The use of cofiring creates an energy that is not as high as pure coal, but emits fewer pollutants and cuts down on the release of previously sequestered carbon. The continuous release of carbon and other greenhouse gasses into the atmosphere leads to an increase in global temperatures. The use of cofiring does not stop this process but decreases the relative emissions of coal power plants.

Use in Developing World

The Legacy Foundation has developed a set of techniques to produce biomass briquettes through artisanal production in rural villages that can be used for heating and cooking. These techniques were recently pioneered by Virunga National Park in eastern Democratic Republic of Congo, following the massive destruction of the mountain gorilla habitat for charcoal.

Pangani, Tanzania, is an area covered in coconut groves. After harvesting the meat of the coconut, the indigenous people would litter the ground with the husks, believing them to be useless. The husks later became a profit center after it was discovered that coconut husks are well suited to be the main ingredient in bio briquettes. This alternative fuel mixture burns incredibly efficiently and leaves little residue, making it a reliable source for cooking in the undeveloped country. The developing world has always relied on the burning biomass due to its low cost and availability anywhere there is organic material. The briquette production only improves upon the ancient practice by increasing the efficiency of pyrolysis.

Two major components of the developing world are China and India. The economies are rapidly increasing due to cheap ways of harnessing electricity and emitting large amounts of carbon dioxide. The Kyoto Protocol attempted to regulate the emissions of the three different worlds, but there were disagreements as to which country should be penalized for emissions based on its previous and future emissions. The United States has been the largest emitter but China has recently become the largest per capita. The United States had emitted a rigorous amount of carbon dioxide during its development and the developing nations argue that they should not be forced to meet the requirements. At the lower end, the undeveloped nations believe that they have little responsibility for what has been done to the carbon dioxide levels. The major use of biomass briquettes in India, is in industrial applications usually to produce steam. A lot of conversions of boilers from FO to biomass briquettes have happened over the past decade. A vast majority of those projects are registered under CDM (Kyoto Protocol), which allows for users to get carbon credits.

The use of biomass briquettes is strongly encouraged by issuing carbon credits. One carbon credit is equal to one free ton of carbon dioxide to be emitted into the atmosphere. India has started to replace charcoal with biomass briquettes in regards to boiler fuel, especially in the southern parts of the country because the biomass briquettes can be created domestically, depending on the availability of land. Therefore, constantly rising fuel prices will be less influential in an economy if sources of fuel can be easily produced domestically. Lehra Fuel Tech Pvt Ltd is approved by Indian Renewable Energy Development Agency (IREDA), is one of the largest briquetting machine manufacturers from Ludhiana, India.

In the African Great Lakes region, work on biomass briquette production has been spearheaded by a number of NGOs with GVEP (Global Village Energy Partnership) taking a lead in promoting briquette products and briquette entrepreneurs in the three Great Lakes countries; namely, Kenya, Uganda and Tanzania. This has been achieved by a five-year EU and Dutch government sponsored project called DEEP EA (Developing Energy Enterprises Project East Africa) . The main feed stock for briquettes in the East African region has mainly been charcoal dust although alternative like sawdust, bagasse, coffee husks and rice husks have also been used.

Use in Developed World

Coal is the largest carbon dioxide emitter per unit area when it comes to electricity generation. It is also the most common ingredient in charcoal. There has been a recent push to replace the burning of fossil fuels with biomass. The replacement of this nonrenewable resource with biological waste would lower the carbon footprint of grill owners and lower the overall pollution of the world. Citizens are also starting to manufacture briquettes at home. The first machines would create briquettes for homeowners out of compressed sawdust, however, current machines allow for briquette production out of any sort of dried biomass.

Arizona has also taken initiative to turn waste biomass into a source of energy. Waste cotton and pecan material used to provide a nesting ground for bugs that would destroy the new crops in the spring. To stop this problem farmers buried the biomass, which quickly led to soil degradation. These materials were discovered to be a very efficient source of energy and took care of issues that had plagued farms.

The United States Department of Energy has financed several projects to test the viability of biomass briquettes on a national scale. The scope of the projects is to increase the efficiency of gasifiers as well as produce plans for production facilities.

Criticism

Biomass is composed of organic materials, therefore, large amounts of land are required to produce the fuel. Critics argue that the use of this land should be utilized for food distribution rather than crop degradation. Also, climate changes may cause a harsh season, where the material extracted will need to be swapped for food rather than energy. The assumption is that the production of biomass decreases the food supply, causing an increase in world hunger by extracting the organic materials such as corn and soybeans for fuel rather than food.

The cost of implementing a new technology such as biomass into the current infrastructure is also high. The fixed costs with the production of biomass briquettes are high due to the new undeveloped technologies that revolve around the extraction, production and storage of the biomass. Technologies regarding extraction of oil and coal have been developing for decades, becoming more efficient with each year. A new undeveloped technology regarding fuel utilization that has no infrastructure built around makes it nearly impossible to compete in the current market.

ADVANTAGES AND DISADVANTAGES OF BIOFUEL

Advantages of Biofuels

1. Cost Benefit: As of now, biofuels cost the same in the market as gasoline does. However, the overall cost benefit of using them is much higher. They are cleaner fuels, which means they produce fewer emissions on burning. Biofuels are adaptable

to current engine designs and perform very well in most conditions. This keeps the engine running for longer, requires less maintenance and brings down overall pollution check costs. With the increased demand of biofuels, they have a potential of becoming cheaper in future as well. So, the use of biofuels will be less of a drain on the wallet.

2. Easy To Source: Gasoline is refined from crude oil, which happens to be a non-renewable resource. Although current reservoirs of gas will sustain for many years, they will end sometime in near future. Biofuels are made from many different sources such as manure, waste from crops and plants grown specifically for the fuel.

3. Renewable: Most of the fossil fuels will expire and end up in smoke one day. Since most of the sources like manure, corn, switchgrass, soyabeans, waste from crops and plants are renewable and are not likely to run out any time soon, making the use of biofuels efficient in nature. These crops can be replanted again and again.

4. Reduce Greenhouse Gases: Fossil fuels, when burnt, produce large amount of greenhouse gases i.e. carbon dioxide in the atmosphere. These greenhouse gases trap sunlight and cause planet to warm. The burning of coal and oil increases the temperature and causes global warming. To reduce the impact of greenhouse gases, people around the world are using biofuels. Studies suggests that biofuels reduces greenhouse gases up to 65 percent.

5. Economic Security: Not every country has large reserves of crude oil. For them, having to import the oil puts a huge dent in the economy. If more people start shifting towards biofuels, a country can reduce its dependance on fossil fuels. More jobs will be created with a growing biofuel industry, which will keep our economy secure.

6. Reduce Dependance on Foreign Oil: While locally grown crops has reduce the nation's dependance on fossil fuels, many experts believe that it will take a long time to solve our energy needs. As prices of crude oil is touching sky high, we need some more alternative energy solutions to reduce our dependance on fossil fuels.

7. Lower Levels of Pollution: Since biofuels can be made from renewable resources, they cause less pollution to the planet. However, that is not the only reason why the use of biofuels is being encouraged. They release lower levels of carbon dioxide and other emissions when burnt. Although the production of biofuels creates carbon dioxide as a byproduct, it is frequently used to grow the plants that will be converted into the fuel. This allows it to become something close to a self sustaining system.

Disadvantages of Biofuels

1. High Cost of Production: Even with all the benefits associated with biofuels, they are quite expensive to produce in the current market. As of now, the interest and capital investment being put into biofuel production is fairly low but it can match demand. If

the demand increases, then increasing the supply will be a long term operation, which will be quite expensive. Such a disadvantage is still preventing the use of biofuels from becoming more popular.

2. Monoculture: Monoculture refers to practice of producing same crops year after year, rather than producing various crops through a farmer's fields over time. While, this might be economically attractive for farmers but growing same crop every year may deprive the soil of nutrients that are put back into the soil through crop rotation.

3. Use of Fertilizers: Biofuels are produced from crops and these crops need fertilizers to grow better. The downside of using fertilizers is that they can have harmful effects on surrounding environment and may cause water pollution. Fertilizers contain nitrogen and phosphorus. They can be washed away from soil to nearby lake, river or pond.

4. Shortage of Food: Biofuels are extracted from plants and crops that have high levels of sugar in them. However, most of these crops are also used as food crops. Even though waste material from plants can be used as raw material, the requirement for such food crops will still exist. It will take up agricultural space from other crops, which can create a number of problems. Even if it does not cause an acute shortage of food, it will definitely put pressure on the current growth of crops. One major worry being faced by people is that the growing use of biofuels may just mean a rise in food prices as well.

5. Industrial Pollution: The carbon footprint of biofuels is less than the traditional forms of fuel when burnt. However, the process with which they are produced makes up for that. Production is largely dependent on lots of water and oil. Large scale industries meant for churning out biofuel are known to emit large amounts of emissions and cause small scale water pollution as well. Unless more efficient means of production are put into place, the overall carbon emission does not get a very big dent in it.

6. Water Use: Large quantities of water are required to irrigate the biofuel crops and it may impose strain on local and regional water resources, if not managed wisely. In order to produce corn based ethanol to meet local demand for biofuels, massive quantities of water are used that could put unsustainable pressure on local water resources.

7. Future Rise in Price: Current technology being employed for the production of biofuels is not as efficient as it should be. Scientists are engaged in developing better means by which we can extract this fuel. However, the cost of research and future installation means that the price of biofuels will see a significant spike. As of now, the prices are comparable with gasoline and are still feasible. Constantly rising prices may make the use of biofuels as harsh on the economy as the rising gas prices are doing right now.

References

- Biofuel, technology: britannica.com, Retrieved 13 July, 2019
- Top-10-uses-for-biofuel: sgbiofuels.com, Retrieved 16 June, 2019
- Biofuels-1, paper3, to-the-points: drishtiias.com, Retrieved 17 February, 2019
- Dubé, marc a, et al. (2007). "acid-catalyzed transesterification of canola oil to biodiesel under single- and two-phase reaction conditions". Energy & fuels 21: 2450–2459. American chemical society. Retrieved on 2007-11-01
- What-is-biodiesel: conserve-energy-future.com, Retrieved 21 June, 2019
- National non-food crops centre. "nnfcc renewable fuels and energy factsheet: anaerobic digestion", retrieved on 2011-02-16
- What-is-biodiesel: conserve-energy-future.com, Retrieved 2 April, 2019
- How-is-biogas-produced, biogas, biogas, about-gas: gasum.com, Retrieved 28 August, 2019
- Advantages-and-disadvantages-of-biofuels: conserve-energy-future.com, Retrieved 1 February, 2019

PERMISSIONS

All chapters in this book are published with permission under the Creative Commons Attribution Share Alike License or equivalent. Every chapter published in this book has been scrutinized by our experts. Their significance has been extensively debated. The topics covered herein carry significant information for a comprehensive understanding. They may even be implemented as practical applications or may be referred to as a beginning point for further studies.

We would like to thank the editorial team for lending their expertise to make the book truly unique. They have played a crucial role in the development of this book. Without their invaluable contributions this book wouldn't have been possible. They have made vital efforts to compile up to date information on the varied aspects of this subject to make this book a valuable addition to the collection of many professionals and students.

This book was conceptualized with the vision of imparting up-to-date and integrated information in this field. To ensure the same, a matchless editorial board was set up. Every individual on the board went through rigorous rounds of assessment to prove their worth. After which they invested a large part of their time researching and compiling the most relevant data for our readers.

The editorial board has been involved in producing this book since its inception. They have spent rigorous hours researching and exploring the diverse topics which have resulted in the successful publishing of this book. They have passed on their knowledge of decades through this book. To expedite this challenging task, the publisher supported the team at every step. A small team of assistant editors was also appointed to further simplify the editing procedure and attain best results for the readers.

Apart from the editorial board, the designing team has also invested a significant amount of their time in understanding the subject and creating the most relevant covers. They scrutinized every image to scout for the most suitable representation of the subject and create an appropriate cover for the book.

The publishing team has been an ardent support to the editorial, designing and production team. Their endless efforts to recruit the best for this project, has resulted in the accomplishment of this book. They are a veteran in the field of academics and their pool of knowledge is as vast as their experience in printing. Their expertise and guidance has proved useful at every step. Their uncompromising quality standards have made this book an exceptional effort. Their encouragement from time to time has been an inspiration for everyone.

The publisher and the editorial board hope that this book will prove to be a valuable piece of knowledge for students, practitioners and scholars across the globe.

INDEX